Operation Sense

Even More Fractions, Decimals, and Percents

STUDENT BOOK

TERC

Mary Jane Schmitt and Myriam Steinback

McGraw Hill Education

Bothell, WA • Chicago, IL • Columbus, OH • New York, NY

TERC
2067 Massachusetts Avenue
Cambridge, Massachusetts 02140

EMPower Research and Development Team
Principal Investigator: Myriam Steinback
Co-Principal Investigator: Mary Jane Schmitt
Research Associate: Martha Merson
Curriculum Developers: Tricia Donovan, Donna Curry

Contributing Author
Andee Rubin

Technical Team
Production and Design Coordinator: Valerie Martin
Copyeditor: Jill Pellarin

Evaluation Team
Brett Consulting Group:
Belle Brett
Marilyn Matzko

EMPower™ was developed at TERC in Cambridge, Massachusetts. This material is based upon work supported by the National Science Foundation under award number ESI-9911410 and by the Education Research Collaborative at TERC. Any opinions, findings, and conclusions or recommendations expressed in this publication are those of the authors and do not necessarily reflect the views of the National Science Foundation.

TERC is a not-for-profit education research and development organization dedicated to improving mathematics, science, and technology teaching and learning.

http://empower.terc.edu

Printed in the United States of America
2 3 4 5 6 7 8 9 QDB 15 14 13 12 11

ISBN: 978-0-07662-094-4
MHID 0-07-662094-8

Contents

Introduction v

Opening the Unit: Operation Sense 1

Lesson 1: Equivalents 7

Lesson 2: Addition—Combining 25

Lesson 3: Subtraction—Take Away, Comparison, and Difference 43

Lesson 4: Multiplication—Repeated Addition and Portions of Amounts 61

Lesson 5: Division—Splitting and Sharing 81

Lesson 6: Division—How Many ____ in ____? 99

Lesson 7: Mixing It Up 115

Closing the Unit: Putting It Together 129

Appendices:

Vocabulary 135

Reflections 137

Introduction

Welcome to EMPower

Students using the EMPower books often find that EMPower's approach to mathematics is different from the approach found in other math books. For some students, it is new to talk about mathematics and to work on math in pairs or groups. The math in the EMPower books will help you connect the math you use in everyday life to the math you learn in your courses.

We asked some students what they thought about EMPower's approach. We thought we would share some of their thoughts with you to help you know what to expect.

> *"It's more hands-on."*
>
> *"More interesting."*
>
> *"I use it in my life."*
>
> *"We learn to work as a team."*
>
> *"Our answers come from each other… [then] we work it out ourselves."*
>
> *"Real-life examples like shopping and money are good."*
>
> *"The lessons are interesting."*
>
> *"I can help my children with their homework."*
>
> *"It makes my brain work."*
>
> *"Math is fun."*

EMPower's goal is to make you think and to give you puzzles you will want to solve. Work hard. Work smart. Think deeply. Ask why.

Using This Book

This book is organized by lessons. Each lesson has the same format.

- The first page explains the lesson and states the purpose of the activity. Look for a question to keep in mind as you work.
- The activity page comes next. You will work on the activities in class, sometimes with a partner or in a group.
- Look for shaded boxes with additional information and ideas to help you get started if you become stuck.

- Practice pages follow the activities. These practices will make sense to you after you have done the activity. The four types of practice pages are

  Practice: provides another chance to see the math from the activity and to use new skills.

  Calculator Practice: offers a chance to thoughtfully explore the calculator, estimate, and look for patterns.

  Extension: presents a challenge with a more difficult problem or a new but related math idea.

  Test Practice: asks a number of multiple-choice questions and one open-ended question.

In the *Appendices* at the end of the book, there is space for you to keep track of what you have learned and to record your thoughts about how you can use the information.

- Use notes, definitions, and drawings to help you remember new words in *Vocabulary*, pages 135–136.
- Answer the *Reflections* questions after each lesson, pages 137–142.

Tips for Success

Where do I begin?

Many people do not know where to begin when they look at their math assignments. If this happens to you, first try to organize your information. Read the problem. Start a drawing to show the situation.

Much of this unit depends on a solid grasp of benchmark fractions, decimals, and percents. By **benchmarks**, we mean friendly, everyday fractions such as $\frac{1}{2}$, $\frac{1}{4}$, $\frac{3}{4}$, $\frac{1}{8}$'s, $\frac{1}{10}$'s, and $\frac{1}{100}$'s; decimals such as 0.1 and 0.01; and the percent equivalents (50%, 25%, 75%, and 10%). Use these benchmarks to reason about the more complicated numbers.

Ask yourself:

Which everyday fraction, decimal, or percent is this number closest to?

People who are strong at math are good estimators. Always take the time to make a reasonable estimate. Benchmarks will come in handy there.

Be flexible. Try to look at the problem in various ways.

Ask yourself:

Will objects, a diagram, or a number line help?

Can I connect this problem to a real-life situation?

Another part of getting organized is figuring out what skills are required.

Ask yourself:

What do I already know? What do I need to find out?

I cannot do it. It seems too hard.

Make the numbers smaller or friendlier. Try to solve the same problem with the benchmark fraction $\frac{1}{2}$.

Ask yourself:

Have I ever seen something like this before? What did I do then?

Take yourself off "automatic." Slow down and reason carefully.

Am I done?

Don't walk away yet. Check your answers to make sure they make sense.

Ask yourself:

Did I answer the question?

Does the answer seem reasonable? Do the conclusions I am drawing seem logical?

Check your math with a calculator. Ask others whether your work makes sense to them.

Practice

Complete as many of the practice pages as you can to sharpen your skills.

© Sean Locke

Opening the Unit: Operation Sense

Where in the newspaper do fractions, decimals, and percents appear?

Before you start a new book, it is good to have an idea where you stand. Do you already know some of the material? Does it look hard? This lesson will help you and your teacher know what will challenge you and what may be new or a review for you in the weeks ahead.

This opening session is an assessment. Its purpose is to welcome you to activities involving operations with **fractions**, **decimals**, and **percents**. Each activity is a window into what you know and what you have yet to learn.

Activity 1: Newspaper and Magazine Search

Look through a magazine or newspaper.

1. Search for examples of fractions, decimals, and percents. Circle them and mark them with an "F" for fraction, "D" for decimal, or "P" for percent.
2. Which form did there seem to be more of when you looked around? Why do you think that is?
3. Pick one of each number form that interests you: a fraction, a decimal, and a percent. For each one, mentally find twice as much, half as much, 10 times as much, and a tenth as much as the original number. Fill in the chart below. If you are not sure of the exact answer, estimate as best you can.

The Number and What It Is About	Twice as Much	Half as Much	10 Times as Much	$\frac{1}{10}$ as Much
The fraction is				
The decimal is				
The percent is				

Activity 2: Four Problems

For each of the math expressions below, draw a picture and write a story.

1. $5\frac{3}{4} + 1\frac{1}{2}$

   a. Picture:

   b. Story:

2. $12.5 - 4.75$

   a. Picture:

   b. Story:

3. $1\frac{1}{2} \times 6$

a. Picture:

b. Story:

4. $5 \div 1\frac{1}{4}$

a. Picture:

b. Story:

Activity 3: Initial Assessment

Your teacher will show you some tasks and ask you to check off how you feel about your ability to solve them. In each case, check off one of the following:

___ Can do ____ Don't know how ____ Not sure

Operation Sense: Even More Fractions, Decimals, and Percents Unit Goals

- Use mathematical symbols, real-world situations, and pictures for each of the four operations with fractions, decimals, and percents.
- Understand
  - Addition as combination;
  - Subtraction as take away, as comparison, and as distance or difference between amounts;
  - Multiplication as repeated addition and as a portion of an amount;
  - Division as sharing and as how many __ in __.
- Use mental math or estimate answers based on benchmark numbers, and then use the scientific calculator to verify or get a precise result.
- Use mathematical properties and the relationships between the four operations to break down problems and check answers.

My Own Goals

LESSON 1

Equivalents

© Steven Lunetta Photography

How do these compare?

Every number can be written in many forms. How many ways do you know to write a number that means the same as 0.8? How are you sure that a number is equal to three-fifths? Numbers may look different but have the same value; they are **equivalent** (like $\frac{1}{2}$ and 50%). Sometimes numbers look almost alike but do not have the same value; they are not equivalent (like 5.0 and 50.).

If you pass a sign on a road too fast or read a headline too quickly, you might think you saw the number 25 when it was actually 0.25. In this lesson, you will need to keep your eyes sharp and pay close attention! Examine the value of the numbers carefully. Give yourself the time to think about the meaning of the numbers you encounter.

Activity 1: Watch Out!

Each of the following scenarios shows someone's reasoning. Try to follow the reasoning, and explain your own way of thinking about each one.

1. Sharon's daughter is in a class of 17 children. She told Sharon, "Today, 12 of the kids in my class were absent." Sharon thought about the number for a minute and said, "That's about two-thirds of the class. Why were so many children absent?"

   a. Explain how Sharon might have reasoned that "about two-thirds of the class" were absent.

   b. Was the number of class members missing actually more than, less than, or equal to two-thirds? Explain.

   c. In what situations is it fine to say "about two-thirds"? When might you need to be more precise? Give examples.

2. Leanne walks into class wondering aloud:

"My son had to find fractions equivalent to $\frac{8}{24}$. I said to him, "You just multiply each part of the fraction by two to get $\frac{16}{24}$." He disagreed, saying, "I think you just divide the **numerator** and **denominator** by eight to get $\frac{1}{3}$."

What do you think? Are Leanne and her son both right? How do you know?

3. Jenny was shopping with her friend Will. The sales tax in their state was 6%. Jenny told Will, "I prefer to think about the sales tax in decimals, so I think of it as 'point six.'"

Do you agree with Jenny? Explain.

Activity 2: Those Zeros and Points

This activity is to be done in pairs. Pick a partner.

1. Your teacher will give you and your partner four cards. Together with your partner, examine these four cards that show three **digits** and a decimal point. Move them around to decide on answers to the questions that follow.

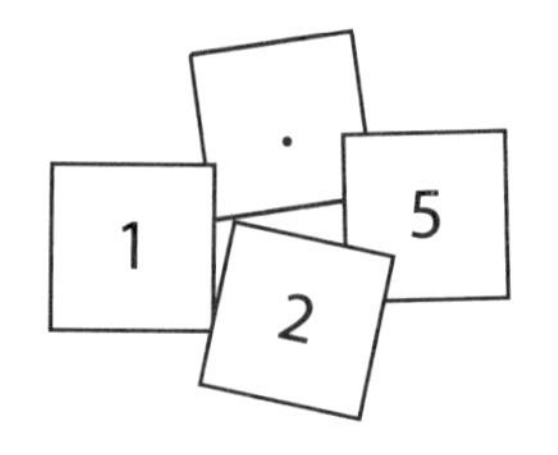

   a. What is the largest possible number you can make with all four cards? Write it down.

   b. What is the smallest possible number you can make with all four cards? Write it down.

   c. Write down a number whose value is in between the numbers you figured out for the first two problems, again using all four cards.

   d. For each of the numbers you found in Problems a, b, and c, think of something that could weigh about that many pounds.

2. Your teacher will give you and your partner four more cards. Examine them and move them around to decide on the answers to the questions that follow.

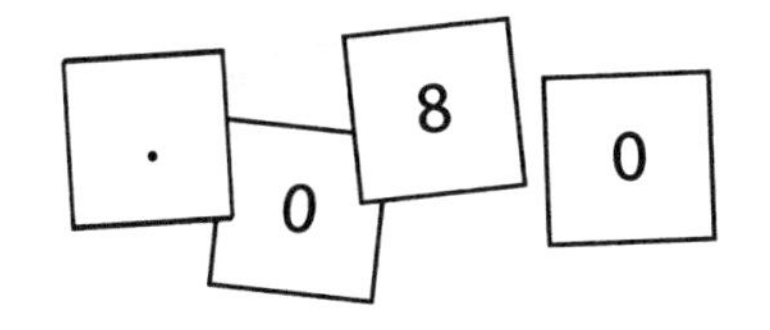

a. What is the largest possible number you can make using all the cards?

b. What is the smallest possible number you can make using all the cards?

c. Make as many numbers as you can, using ALL the cards each time. List all your numbers below.

d. Pick two of the numbers you listed for Problem c. Use words to write out each number.

e. Order the numbers you listed in Problem c from smallest to largest, and list them below.

f. Review your list above. Cross out any zeros that you think are not necessary.

Activity 3: Two Ways to See It

A fraction can be interpreted in more than one way. Take, for example, the fraction $\frac{3}{5}$.

One way to see $\frac{3}{5}$ is as 3 parts out of 5 total parts in the whole.

1. Make a drawing showing either "I ate $\frac{3}{5}$ of the pack of gum" or "I spend $\frac{3}{5}$ of my salary on rent."

Another way to see $\frac{3}{5}$ is as $3 \div 5$, or 3 divided by 5.

2. Make a drawing showing either " I had 3 dollars and shared them with 5 people" or "If it costs $3 for 5 bottles of water, how much does one bottle cost?"

3. Explain both ways you can interpret $\frac{9}{12}$: as parts of a whole and as a division problem. Make a drawing, give an example, or write a story that shows what you mean.

a. One way I see $\frac{9}{12}$ is . . .

b. Another way I see $\frac{9}{12}$ is . . .

Activity 4: Target 1

Play this game with a partner. You may use either a pair of 10-sided dice or a deck of 1–9 cards. Roll two dice. (If a zero comes up, roll again).

Or, randomly pick two cards from a shuffled deck of nine cards, each marked 1 through 9. Use one of the numbers as the numerator of a fraction and the other as the denominator. The objective is to find a fraction as close to 1 as possible.

For example, you roll a 7 and a 2. Your fraction could be $\frac{2}{7}$ or $\frac{7}{2}$. The fraction $\frac{2}{7}$ is closer to 1 than $\frac{7}{2}$.

Your partner rolls a 5 and a 9, and chooses to write $\frac{5}{9}$.

Which is closer to 1: $\frac{2}{7}$ or $\frac{5}{9}$? Explain your thinking, and make sure you and your partner both agree. In the example above, your partner, with $\frac{5}{9}$, wins!

Play six rounds. Circle the winner in each round.

You	Partner
Explain:	

You	Partner
Explain:	

You	Partner

Explain:

You	Partner

Explain:

You	Partner

Explain:

You	Partner

Explain:

Practice: Percent Names

Find the percent name for each of the following fractions or decimals. Try to figure this out in your head first. Use a calculator only when necessary.

1. $\frac{3}{5}$

2. $\frac{6}{10}$

3. $\frac{60}{100}$

4. 0.06

5. 0.60

6. $\frac{3}{8}$

7. 0.625

8. $\frac{1}{16}$

9. $\frac{85}{100}$

10. $\frac{8}{10}$

11. Complete the following sentence:

    In my own life, I find that I use (circle one) fractions, decimals, percents more often. Explain, using an example.

Practice: When Can I Ignore the Zero?

In each of the following problems, decide when you can and can't ignore the zeros. An example has been done for you.

Most cigarettes in the U.S. market contain 10 milligrams (mg) or more of nicotine.

10 ____ I can ignore the zero. _x_ I can't ignore the zero.

Why?

Without the 0, the number would be 1.

1. The U.S. debt on December 13, 2005, was 8.140 trillion dollars.

2005 ____ I can ignore the zeros. ___ I can't ignore the zeros.

8.140 ____ I can ignore the zero. ___ I can't ignore the zero.

Why?

2. Male giraffes weigh between 2,400 and 3,600 pounds.

2,400 ____ I can ignore the zeros. ___ I can't ignore the zeros.

3,600 ____ I can ignore the zeros. ___ I can't ignore the zeros.

Why?

3. One kilometer (km) is approximately equal to 0.62 miles.

0.62 ____ I can ignore the zero. ___ I can't ignore the zero.

Why?

4. The price for downloading a song is $0.88.

$0.88 ____ I can ignore the zero. ___ I can't ignore the zero.

Why?

5. One meter (m) is 1.09361 yards.

1.09361 ____ I can ignore the zero. ___ I can't ignore the zero.

Why?

6. Summarize your work by completing the following sentences:

a. Sometimes zeros are essential to the value of a number. You cannot ignore the zeros when…

b. Sometimes zeros are not essential to the value of a number. You can ignore the zeros when…

Calculator Practice: Fraction-Decimal-Percent Conversion

Complete this equivalence chart to practice making **conversions** among fractions, decimals, and percents. If the numbers are easy, do them in your head. When the numbers get complicated, use a calculator. Or use a paper-and-pencil method. Mark your answers with an H for "in my head," C for "calculator," or P for "paper and pencil."

An example has been done for you.

	Fraction	Decimal	Percent
	$\frac{7}{8}$	0.875	87.5%
1.	$\frac{105}{100}$		
2.			35%
3.		1.5	
4.	$\frac{19}{12}$		
5.			108%
6.		0.59	
7.	$\frac{16}{64}$		
8.	$\frac{5}{52}$		
9.			2%

10. What percent of the calculations did you do in your head? With a calculator? With paper and pencil?

Extension: Just How Many Drugs...

The average retail prescription prices are shown in the following graph:

Retail Spending on Prescription Drugs in the U.S.

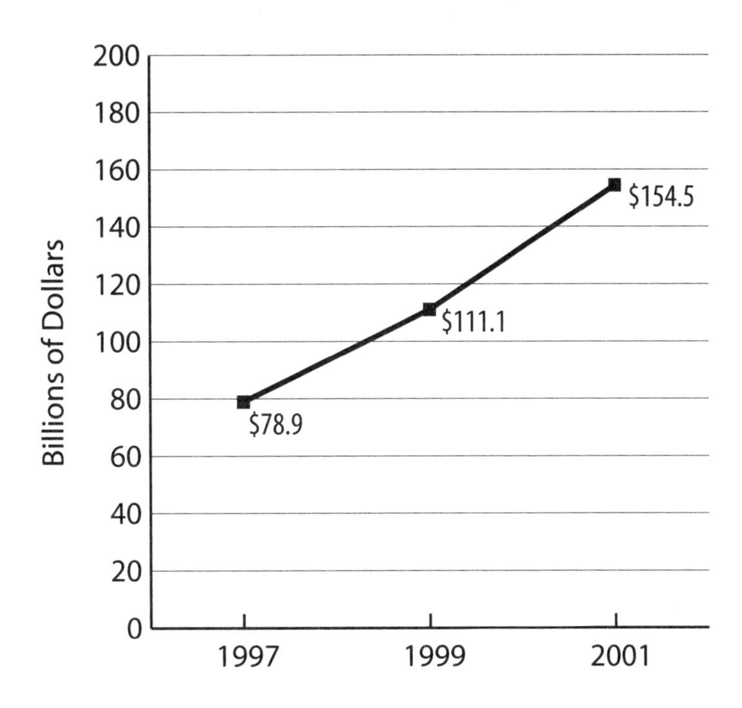

Source: Prescription Drug Expenditures in 2001: Another Year of Escalating Costs. A Report by the National Institute of Health Care Management Research and Educational Foundation, Washington, D.C., 2002

1. According to the graph, about what percent increase was there between 1997 and 2001 in retail prescription spending? Explain.

The most popular drugs by sales, as reported in 2001, were as follows:

Lipitor®: $4.5 billion in sales

Prilosec®: $4.0 billion in sales

Prevacid®: $3.2 billion in sales

Zocor®: $2.7 billion in sales

2. Lipitor sales account for about what portion of the four most popular drug sales reported? Explain.

Test Practice

For each question, choose the best answer.

1. Gloria puts in a 5-day, 40-hour work week at her hospital job. This year she is released for 3 hours, twice a week, to take a college course. Which of the following best describes how much release time she is getting for education?

(1) $\frac{1}{4}$ a day total per week

(2) $\frac{1}{3}$ a day total per week

(3) $\frac{1}{2}$ a day total per week

(4) $\frac{2}{3}$ a day total per week

(5) $\frac{3}{4}$ a day total per week

2. What is the smallest number you can make using the numbers 0, 1, 0, 0, and a decimal point?

(1) 10.00

(2) 1.000

(3) .0010

(4) 100.0

(5) .0001

3. Shelly is talking about the cost of renting a car during the holiday season in Miami. He says, "I spent 60% less when I rented a car during the off-season." In fractions, how much less did Shelly spend?

(1) $\frac{1}{6}$ less

(2) $\frac{1}{3}$ less

(3) $\frac{1}{5}$ less

(4) $\frac{3}{5}$ less

(5) $\frac{3}{4}$ less

4. One way to decide whether $\frac{5}{8}$ or $\frac{5}{12}$ is closer to $\frac{1}{2}$ is to

(1) See that $\frac{5}{12}$ is $\frac{1}{12}$ less than $\frac{1}{2}$, while $\frac{5}{8}$ is $\frac{1}{8}$ more than $\frac{1}{2}$, so $\frac{5}{12}$ is closer to 1, since $\frac{1}{12}$ is less than $\frac{1}{8}$.

(2) See that $\frac{5}{12}$ is more than $\frac{1}{2}$, while $\frac{5}{8}$ is less than $\frac{1}{2}$, so $\frac{5}{12}$ is closer to 1.

(3) Divide 8 by 5 ($8 \div 5$) and 12 by 5 ($12 \div 5$), and compare the two answers.

(4) Change them to 5.8% and 5.12%, and compare.

(5) Change them to 8% and 12%, and compare.

5. Zach made 50 baskets out of 110 free throws that he attempted. What is the benchmark fraction that best describes his performance?

(1) $\frac{1}{4}$

(2) $\frac{1}{3}$

(3) $\frac{1}{2}$

(4) $\frac{2}{3}$

(5) $\frac{3}{4}$

6. In a box of a dozen truffles, eight were missing. What benchmark fraction best describes the number of missing truffles?

LESSON 2

Addition—Combining

© Steven Lunetta Photography

How would you add these amounts?

Going to New York? Your flight will take $\frac{2}{3}$ of an hour. Allow $\frac{1}{2}$ an hour for a taxi ride from the airport. So, how much time will your trip take in all? Knowing how to do fraction addition comes in handy when you combine measurements such as these.

Whenever you are combining two or more amounts, you are adding. In this lesson, you will consider problems that will require knowing how to add fractions, decimals, and percents. Be careful! Fraction addition is similar in some ways to adding whole number amounts, but there are differences. You have to keep track of both parts of the fraction. At the end of this lesson, make a list for yourself of what to remember when adding fractions or decimals.

Activity 1: Watch Out!

Each of the following scenarios shows someone's reasoning. Try to follow the reasoning, and give your own way of thinking about each problem.

1. Luis says to his friend, "I borrowed some money from Pablo. Last month I paid him back 20% of it. Tomorrow I will pay him another 15%, and I plan to pay him the remaining 75% next month."

   Does Luis's plan make sense? Explain. Use a drawing if it's helpful.

2. Elena works at the deli counter in the local grocery store. She serves cold cuts, cheeses, and a variety of salads, all of which she weighs and prices. Her last customer bought 0.8 lb. of turkey breast, 1.2 lbs. of corned beef, and 0.4 lb. of pastrami. Elena said, "Wow! That's a lot of meat—24 lbs.!"

   Do you agree with Elena's statement? Explain.

3. Anita is working on a problem, adding the fractions $\frac{1}{4} + \frac{2}{3}$. She knows something about fractions and reasons in the following way:

$1 + 2 = 3$

$4 + 3 = 7$

"So my answer is $\frac{3}{7}$!" she proclaims.

a. Is Anita's method of reasoning correct? Explain. Use a drawing if it's helpful.

b. Terry is trying to help Anita with her problem. He says, "I have three strips of paper of equal lengths. I folded one into thirds, one into fourths, and one into twelfths. It made it easy to see the answer." Use strips of paper to show what Terry did.

c. Paul says, "I solved the problem using decimals. This method is a little messy, but I think I found the answer." What do you think Paul did?

Activity 2: How Many Is That?

Select three of the following problems and solve them with your partner.

Then, choose one of the problems you solved. On a large piece of paper, show how you solved it, using a diagram and also a math statement or statements.

1. I have $\frac{2}{3}$ of a pound of ground beef. I have a recipe that calls for two pounds of beef. At my supermarket, beef comes in packages of $\frac{1}{4}$ pound.

   a. How many packages do I need to buy to end up with two pounds of ground beef?

   b. How much meat will I have left over? Use folded strips of paper if that is helpful.

2. The football game started at 1:30 p.m. and the first half lasted 75 minutes. The announcer said, "We'll take half an hour for halftime." When did the second half of the game start?

3. I'm worried about heating costs this winter. My bill for the winter last year (November through April, six months) was $1,250. I heard on the radio that heating costs are going up 30% for this year. What will I expect to pay for fuel this winter?

4. The Diner's Club card was the first credit card in the United States, created in 1950. By 1958, it had competition from the American Express card. Now, there are more than 720 different credit cards used at more than 4 million locations in the United States and 11 million more locations around the world.

The average credit card debt per household went from $2,985 in 1990 to $7,564 in 1999. In 2004, that debt was up to $9,312.

a. Joe said, "The average credit card debt increased more than 200% from 1990 to 1999." Do you agree? Explain.

b. Angie replied, "I think you're looking at the wrong numbers. It was from 1990 to 2004 that it increased more than 200%." Do you agree? Explain.

Activity 3: Target 10

For each round, your teacher will toss a die six times and call out the number that comes up.

Each time the die is tossed, fill in the number on one space of the grid for Round 1, either before or after the decimal point. Do this before the die is tossed again.

The space you fill can be anywhere **above** the double line. At the end of six tosses, all six spaces must be filled.

Now add all your numbers. If they add up to more than 10, you lose. The person with a total closest to, but not more than, 10 wins the round.

Play six rounds.

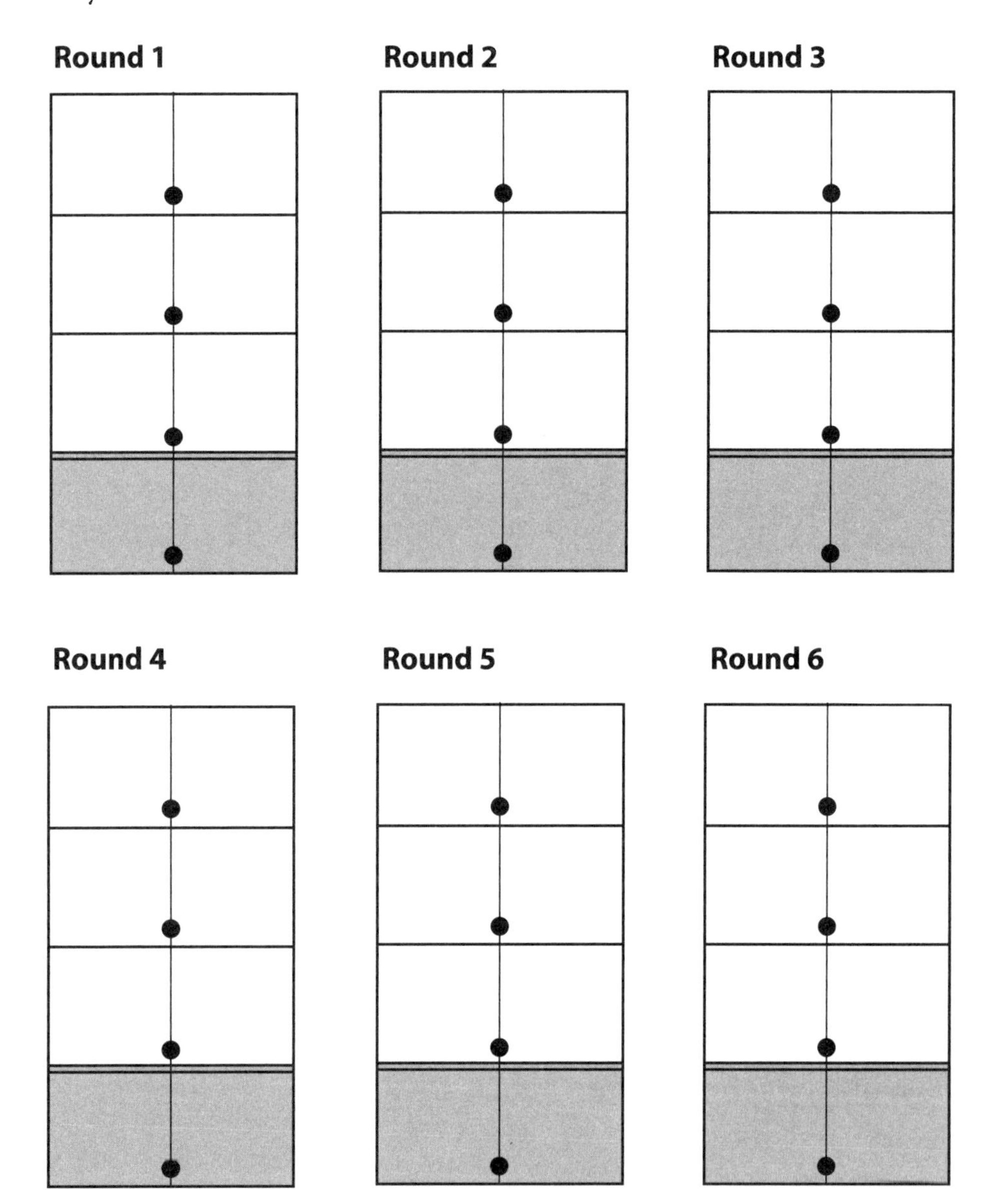

Activity 4: Combining Pattern Blocks

You will need a pile of pattern blocks for this activity.

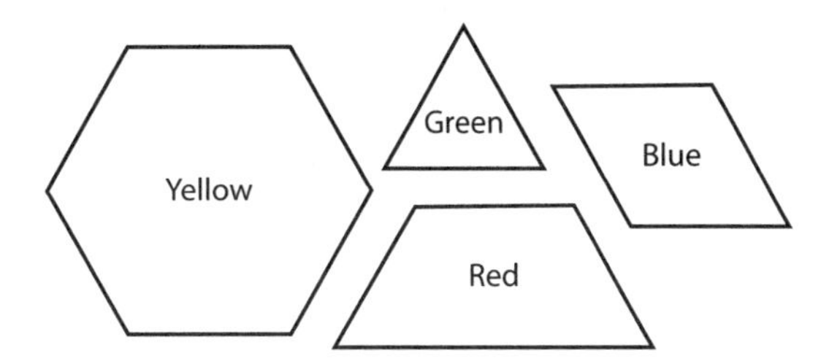

1. Using the (yellow) **hexagon** as the whole (1), find the *fractional* value of

   a. The (green) **triangle**.

   b. The (red) **trapezoid**.

   c. The (blue) **parallelogram**.

2. Create as many combinations of pattern blocks as possible to make the whole. (Use the yellow, red, green, and blue blocks.) Then write an **equation** (a math sentence), using fractions for each combination of pattern blocks you created.

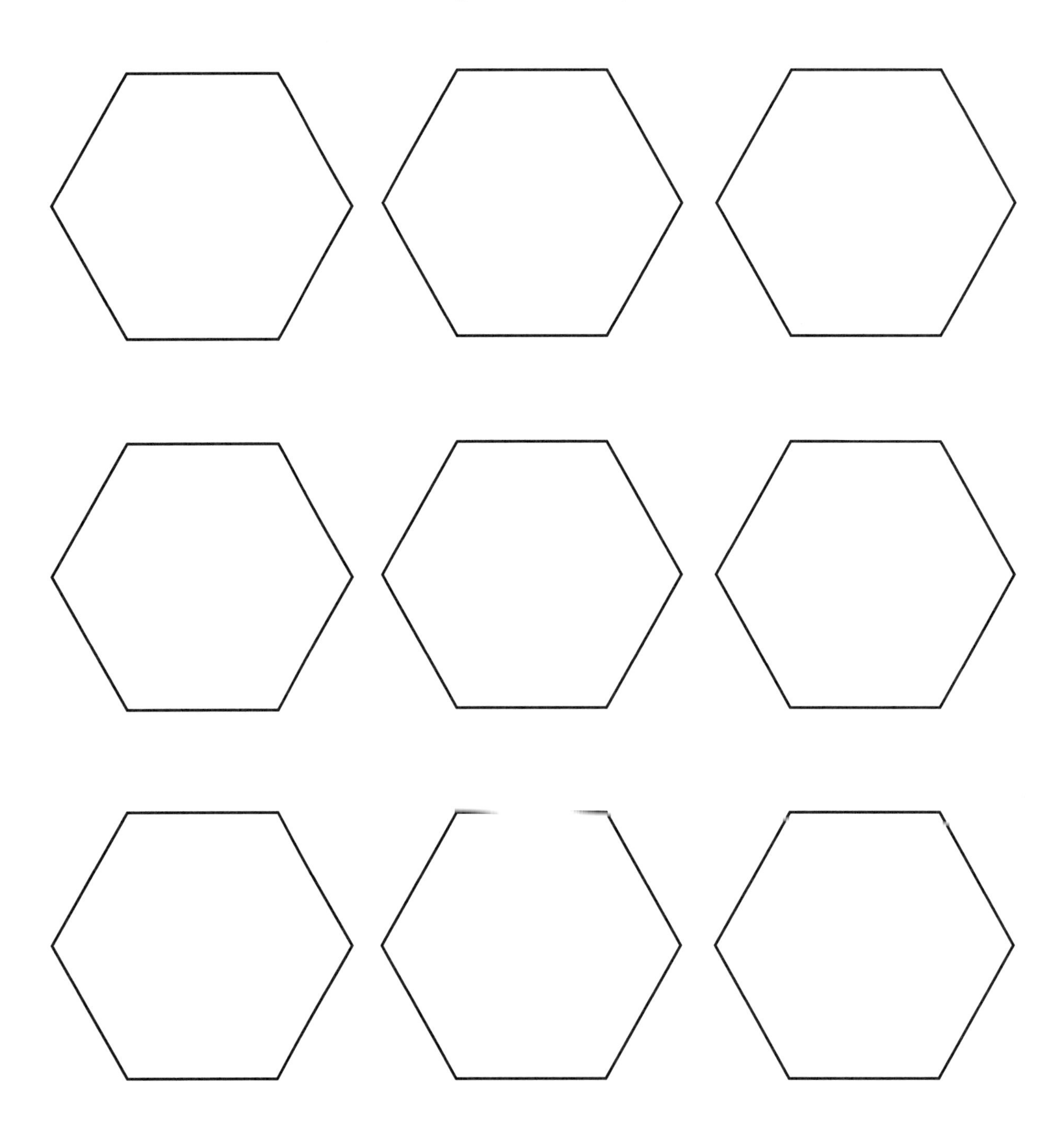

3. Create as many combinations of pattern blocks as possible to make $\frac{1}{2}$. Then write an equation (a math sentence), using fractions for each combination of pattern blocks you created.

4. Which is more, five triangles or three trapezoids?

a. Show by drawing a picture.

b. Write the math equation.

5. Which is more, two parallelograms or four triangles?

a. Show by drawing a picture.

b. Write the math equation.

6. Choose pattern blocks that combined show $\frac{5}{6}$. Do this using two different combinations of pattern blocks. Make drawings of your two combinations, and write the combinations using fractions.

a. One way to show $\frac{5}{6}$:

b. Another way to show $\frac{5}{6}$:

Practice: Where Is the Point?

Place the decimal point where you think it goes in the underlined number.

1. $1.206 + 3.02 = \underline{4226}$

2. $\underline{1206} + 3.02 = 123.62$

3. $1.206 + \underline{302} = 1.508$

4. $1.206 + \underline{302} = 31.406$

5. $0.1206 + \underline{302} = 3.1406$

6. Make a note for yourself: When adding decimal numbers, remember…

7. .708 lb. + 4.02 lb. =

   a. 11.10 lb.

   b. 1,110 lb.

   c. 74.82 lb.

   d. 4.728 lb.

   e. 0.1110 lb.

Practice: Right or Wrong?

Decide whether the addition shown in each problem is right or wrong. Explain your reasoning.

1. $\frac{2}{11} + \frac{7}{11} = \frac{9}{11}$

___ Right ___ Wrong

Explain.

2. $\frac{1}{5} + \frac{3}{7} = \frac{4}{12}$

___ Right ___ Wrong

Explain.

3. 1.206 + 3.02 = 1.508

___ Right ___ Wrong

Explain.

4. 30.3% + 3.03% = 33.33%

___ Right ___ Wrong

Explain.

Practice: Are We There Yet?

1. Elena and Ben were driving from Chicago to Minneapolis with their son, Sam. They started at 8 a.m., drove for 4 hours, and stopped for lunch at a rest area for $\frac{3}{4}$ of an hour. Then they continued driving, and after $3\frac{1}{2}$ hours, Sam asked, "Are we there yet?"

   a. How long had they been on the road? Explain. If a number line is helpful, draw one and use it.

   b. What time was it when Sam asked his question? Explain.

2. Jeremy loves checking the odometer in his dad's car. On a long trip, he noted that when they started, the odometer read 108.9. His dad said that the odometer would read 319.5 when they got to their destination. At one point, Jeremy asked, "Are we there yet?" His Dad responded by telling him that they had driven 210 miles.

   a. Are they there yet? How do you know?

   b. How close to their destination are they? Explain.

Calculator Practice: Adding Fractions and Decimals

Estimate the answer first. Then, use the calculator to check your estimate and find the exact answer. An example has been done for you.

Problem	Your Estimate	Calculator Answer
$\frac{1}{4} + 10.8$	A little more than 11, because I know that 10.8 is a little more than $10\frac{3}{4}$; and if we add $\frac{1}{4}$, the answer is 11 plus a little more.	$\frac{1}{4} = 0.25$ $0.25 + 10.8 = 11.05$
1. $\frac{1}{5} + 0.5$		
2. $1.75 + \frac{1}{8}$		
3. $0.08 + \frac{2}{3}$		
4. $5.33 + \frac{1}{10}$		
5. $16.67 + 2\frac{1}{3}$		
6. $\frac{12}{72} + 0.375$		

Extension: Measurement Conversions

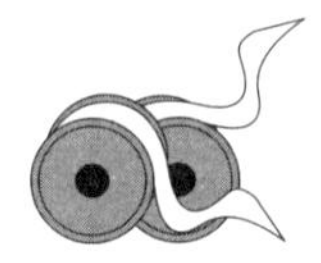

1. Serena is wrapping birthday presents. For one of them she needs 1 foot of ribbon, and for the other one she needs $\frac{1}{4}$-yard of ribbon. How much ribbon does Serena need? Express the answer in yards.

   (1 yard = 3 feet; 1 foot = 12 inches)

2. Chris is a chef at a local restaurant. She is making two desserts that call for milk and would like help figuring out how much milk she needs. For one, she needs $\frac{1}{2}$ a pint of milk, and for the other, she needs $\frac{1}{2}$ a gallon of milk. How much milk does Chris need all together for her desserts? Express the answer in gallons. What is the answer in quarts?

   (2 cups = 1 pint; 2 pints = 1 quart; 4 quarts = 1 gallon)

Test Practice

For each question, choose the best answer.

1. Emily and her sister Jane were eating a pizza. Jane ate $\frac{1}{3}$ of the pizza and Emily ate $\frac{1}{2}$. How much of the pizza did they eat together?
   (1) The whole pizza
   (2) $\frac{2}{5}$ of the pizza
   (3) $\frac{5}{6}$ of the pizza
   (4) Less than $\frac{1}{2}$ of the pizza
   (5) Less than $\frac{1}{3}$ of the pizza

2. Add: 5.84 + 8 + 12.97
   (1) .1961
   (2) 2.681
   (3) 18.89
   (4) 19.61
   (5) 26.81

3. If the hexagon pattern block is the whole, then which arrangement of shapes shows $\frac{2}{3} + \frac{1}{2}$?

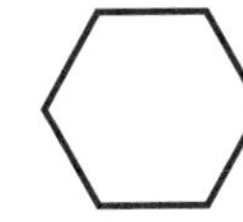

   (1)  
   (2) 
   (3) 
   (4) 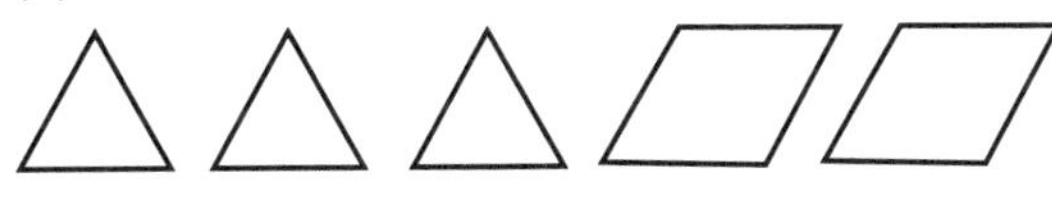
   (5)

4. Find the missing number: 5.6 + 0.3 + ____ = 10
   (1) .14
   (2) 1.4
   (3) 4.1
   (4) 15.9
   (5) 69

5. In 1989, salads peaked when they accounted for 10% of main courses in all restaurant meals. In 2005, it was $5\frac{1}{2}$%. How much higher was the percent of salads as the main course in restaurant meals in 1989 than in 2005?
   (1) 3%
   (2) $4\frac{1}{2}$%
   (3) 5%
   (4) $5\frac{1}{2}$%
   (5) $15\frac{1}{2}$%

6. 170.9 + 3.10 =

LESSON 3

© Pierre-Emmanuel Turcotte

Subtraction—Take Away, Comparison, and Difference

Where is the subtraction?

What picture comes to mind from this problem: 1,000 – 350?

You might interpret it as something to do with a "take-away" situation and see it in terms of money: "If I have $1,000 to start with, and I spend, or take away, $350, that leaves me with $650." Or, you might think of the problem in terms of a **number line** and ask, "What is the distance between the two amounts, 1,000 miles and 350 miles? The answer is 650 miles." Finally, you could ask, "How much heavier is a 1,000-pound crate than a 350-pound crate?' These are three very different, yet correct, pictures of that simple subtraction problem. In all cases, addition can be used to check the result: 650 + 350 = 1,000.

In this lesson, you will be challenged to picture subtraction with fraction and decimal numbers. Be flexible in your thinking and creative with your drawings. Ask yourself: Can I see this as a "take-away" situation, as the distance between two numbers, or as how much more one amount is than another?

Activity 1: Show Me...

1. Show me $15\frac{1}{2} - 5\frac{7}{8}$

   a. Make a drawing to communicate what the problem and its solution mean.

   b. Write a story for which you could use $15\frac{1}{2} - 5\frac{7}{8}$ as a way to find the solution.

   c. How is $15\frac{1}{2} - 5\frac{7}{8}$ different from $15\frac{7}{8} - 5\frac{1}{2}$?

2. Show me 2.3 – 1.75

a. Make a drawing to communicate what the problem and its solution mean.

b. Write a story for which you could use 2.3 – 1.75 as a way to find the solution.

c. How is 2.3 – 1.75 different from 2.75 – 1.3?

Activity 2: Watch Out!

Each of the following scenarios shows some people's reasoning. Try to follow the reasoning, and give your own way of thinking about each one.

Scenario 1

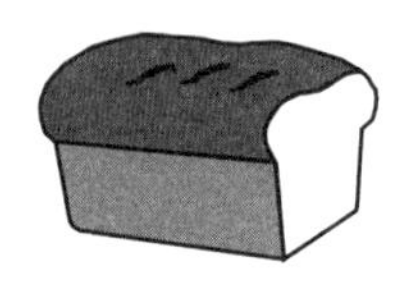

Alicia and Meredith each buy identical loaves of bread. Alicia cut hers into 8 slices and ate 3 of them. Meredith cut hers into 10 slices and ate 4 of them. They were each claiming they ate more than the other.

Alicia said, "I ate more than you because I ate almost half my loaf." Meredith countered, "No, I ate almost half my loaf, and I ate one more slice than you!"

Who do you think ate the most? Explain.

Scenario 2

Jerry is working on a subtraction of fractions problem:

$\frac{7}{8} - \frac{3}{4}$

He knows something about fractions and reasons in this way:

$7 - 3 = 4$

$8 - 4 = 4$

"So, my answer is $\frac{4}{4}$, or 1!"

Did Jerry figure out the problem correctly? Explain. Use a drawing if it's helpful.

Scenario 3

Arthur says, "I used **common denominators** to solve Jerry's problem."

Show what you think Arthur did. Use paper strips (as in *Lesson 2*) if that is helpful.

Scenario 4

Larry announces, "I solved the problem by using decimals. It's a little messy, but I think I found the answer."

What do you think Larry did?

Scenario 5

This chart was handed to a reporter covering the Falmouth 10K Road Race.

	Male Winning Times (min.)	Female Winning Times (min.)
Youth (under 16)	36.5	40
Adult (16–40)	29.54	35
Seniors (40 plus)	40.1	46

The reporter quickly typed and turned in this story:

Mighty Fast Females

On Saturday, there was a huge upset when female runners outdid male runners in every age category. Fifteen-year-old SueAnne Evans beat the winning boy, Harvey Yazijian, by 32.5 minutes. Mia Calla, age 23, beat Sam Paley by 29.19 minutes. And the biggest upset of all was when 45-year-old Lucy Marza completed the race 35.5 minutes before Mark LaRue.

Do you agree with the reporter's math? Explain.

Scenario 6

The editor sent an angry note to this reporter asking him to fix his errors and resubmit. Rewrite the story for the reporter.

Activity 3: Fractions and Decimals on the Line

1. Place the following numbers on the number line.

   0.4 $\frac{3}{5}$ 0.35 $\frac{1}{4}$

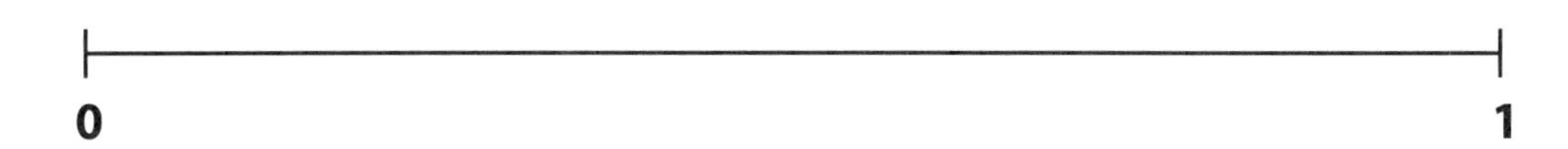

2. a. What is the difference between the smallest and the largest numbers?

   b. Show that difference on the number line.

   c. Did you add or subtract to find the difference?

3. Show the answer to $\frac{3}{5} - 0.4$ on the number line.

Activity 4: Target 1, $\frac{1}{2}$, $\frac{1}{4}$, and 0

For Rounds 1 and 2 of this game, you toss a 10–sided die three times. If you do not have a 10-sided die, use a set of 0–9 cards, replacing each card before you choose another.

For Round 3, use the fraction cards that your teacher will give you.

Round 1

Write the three **digits** from your die tosses or cards here:

_______ _______ _______

Use a decimal point and the three digits to

1. Write the decimal number that has a value closest to 1.

2. Write the decimal number that has a value closest to $\frac{1}{2}$.

3. Write the decimal number that has a value closest to $\frac{1}{4}$.

4. What is the difference between the smallest and the largest numbers you wrote?

5. Compare your answers with a partner. Come to an agreement on the best answers. What did you notice?

Round 2

Write the three digits from your die tosses or cards here:

_______ _______ _______

Use a decimal point and the three digits to

1. Write the decimal number that has a value closest to 1.

2. Write the decimal number that has a value closest to $\frac{1}{2}$.

3. Write the decimal number that has a value closest to $\frac{1}{4}$.

4. What is the difference between the smallest and the largest number you wrote?

5. Compare your answers with a partner. Come to an agreement on the best answers. What did you notice?

Round 3

Examine the fraction cards.

1. Find three pairs of fractions whose value have a difference closest to 0. Write the fraction pairs below. For each one, explain how you know that their difference is very close to 0.

   ________ ________

   ________ ________

   ________ ________

2. Compare your answers with a partner. Come to an agreement on the best answers. What did you notice?

Practice: Where Is the Point?

Do these problems in your head. No calculators or pencils and paper!

Place the decimal point where you think it goes in the underlined number.

1. 11.206 – 3.02 = <u>8186</u>
2. <u>1266</u> – 3.02 = 123.58
3. 1.6 – <u>302</u> = 1.298
4. 31.104 – <u>302</u> = 0.904
5. 1.206 + <u>302</u> = 303.206

Select the best answer for each of these problems.

6. 3.02 – 1.8 =
   - **a.** 0.122
   - **b.** 1.22
   - **c.** 1.4
   - **d.** 1.78
   - **e.** 2.6

7. 0.7 – 0.25
   - **a.** 0.045
   - **b.** 0.18
   - **c.** 0.45
   - **d.** 1.8
   - **e.** 4.5

Practice: Where on the Line?

1. Refer to the table with data about numbers of people (in millions) diagnosed with diabetes in the United States between 1980 and 2004.

   a. Mark the lowest and and the highest number of people diagnosed with diabetes on the number line.

   b. How many more people were diagnosed with diabetes in 2004 than in 1980?

   c. During which year did the number of people diagnosed with diabetes in 1980 number increase by 100%? How do you know? Show the increase on the number line.

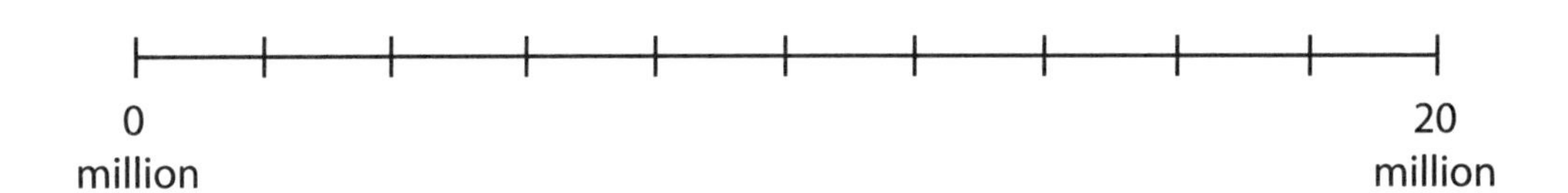

Prevalence of Diabetes*

Number (in Millions) of Persons with Diagnosed Diabetes, United States, 1980–2004

Year	Number
1980	5.8
1981	5.8
1982	5.8
1983	5.9
1984	6.1
1985	6.4
1986	6.6
1987	6.6
1988	6.5
1989	6.4
1990	6.7
1991	7.0
1992	7.6
1993	7.8
1994	8.3
1995	8.2
1996	8.5
1997	10.3
1998	10.5
1999	11.1
2000	12.0
2001	12.9
2002	13.6
2003	14.3
2004	14.7

*Source: http://www.cdc.gov/diabetes/statistics/prev/national/tablepersons.htm

2. The monthly mean level of cholesterol for men and women from August to December 2004 is described below.

Month	Female	Male
Aug-04	114.0	110.4
Sep-04	113.5	110.2
Oct-04	113.1	109.9
Nov-04	112.7	110.2
Dec-04	112.5	110.9

Source: http://www.questidagnostics.com/brand/business/healthtrends/hearthealth/data.html#bygender

a. Locate the highest and lowest mean cholesterol levels for men on the number line. Find and show the difference.

b. On the same number line, locate the highest and lowest mean cholesterol levels for women. Find and show the difference.

c. Whose mean cholesterol levels changed the most, the men's or the women's? Explain.

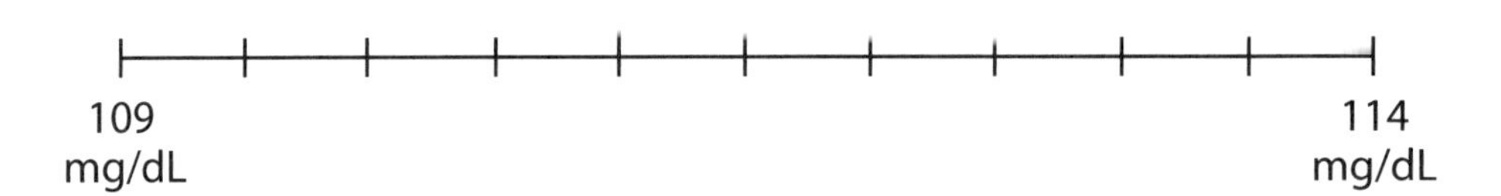

Calculator Practice: Subtracting Fractions and Decimals

Estimate the answer first. Then use the calculator to check or get the answer. An example has been done for you.

Problem	Your Estimate	Calculator Answer
$10.736 - \frac{1}{2}$	A little more than 10	$\frac{1}{2} = 0.5$ $10.736 - 0.5 =$ 10.236
1. $0.5 - \frac{1}{5}$		
2. $1.75 - \frac{1}{8}$		
3. $\frac{2}{3} - 0.08$		
4. $5.33 - \frac{1}{10}$		
5. $16.67 - 2\frac{2}{3}$		
6. $0.375 - \frac{12}{72}$		

Extension: How Much More?

The following are the ingredients for two different recipes for a dozen muffins.

Applesauce Spice Muffins	Apple Mini-Muffins
$1\frac{1}{2}$ cups all-purpose flour	$\frac{1}{4}$ cup walnuts
$1\frac{1}{2}$ teaspoons baking powder	$1\frac{1}{2}$ teaspoons baking powder
$\frac{1}{2}$ teaspoon baking soda	$\frac{1}{4}$ teaspoon salt
$\frac{1}{2}$ teaspoon cinnamon	$\frac{1}{2}$ Granny Smith or Golden Delicious apple
$\frac{1}{2}$ teaspoon ground allspice	3 tablespoons unsalted butter
$\frac{1}{4}$ teaspoon freshly grated nutmeg	$\frac{1}{4}$ cup packed brown sugar
$\frac{1}{4}$ teaspoon salt	1 large egg
2 large eggs	$\frac{1}{2}$ teaspoon cinnamon
1 cup packed light brown sugar	1 cup all-purpose flour
1 stick ($\frac{1}{2}$ cup) plus 3 tablespoons unsalted butter, melted	2 tablespoons unpasteurized apple cider
1 cup unsweetened applesauce	
1 cup pecans or walnuts ($3\frac{1}{2}$ oz), coarsely chopped	

1. How much more flour is in the applesauce spice muffins than in the apple mini-muffins?

2. What other differences do you notice in ingredients between the two muffin recipes?

3. Normal body temperature for an adult is 98.6 degrees Fahrenheit. John spiked a fever when he had the flu. The thermometer read 102.4°. How many degrees above normal was John's temperature?

   a. Explain, using subtraction.

   b. Explain, using addition.

   c. How are Problems a and b related?

4. Alyssa is trying to lose weight. Her target weight is 135 lbs. On week three, she is at 172.5 lbs. How many more pounds does she need to lose to reach her target weight?

5. In a football game, one team had to run from the 45.5-yard line to the 32-yard line in one play. When the other team was on the offensive, they had to run from the 32.5-yard line to the 45-yard line. Which team had to run farther? Explain.

Test Practice

For each question, choose the best answer.

1. $12\frac{1}{8} - 1\frac{1}{2} =$

   (1) $10\frac{1}{2}$

   (2) $10\frac{5}{8}$

   (3) $10\frac{1}{6}$

   (4) $11\frac{1}{8}$

   (5) $11\frac{1}{6}$

2. How much taller is Joey, 5 ft. 6.5 in., than Carole, 5 ft. 3.75 in.?

   (1) 1.5 in.

   (2) 1.75 in.

   (3) 2 in.

   (4) 2.75 in.

   (5) 3.25 in.

3. How much more is 2 than $\frac{8}{5}$?

   (1) $\frac{6}{5}$

   (2) $\frac{3}{5}$

   (3) $\frac{2}{5}$

   (4) $\frac{1}{5}$

   (5) 1

4. Albert has $\frac{1}{3}$ less marbles in his collection than Ben has. Ben has 24 marbles in his collection. How many marbles are in Albert's collection?

   (1) 3

   (2) 8

   (3) 13

   (4) 16

   (5) $23\frac{1}{3}$

5. Consider these numbers: $\frac{4}{5}$ .45 .04

   What is the difference between the largest and smallest number?

   (1) 0

   (2) .41

   (3) .76

   (4) .80

   (5) 1

6. $130.091 - 2.01 =$

LESSON 4

Multiplication—Repeated Addition and Portions of Amounts

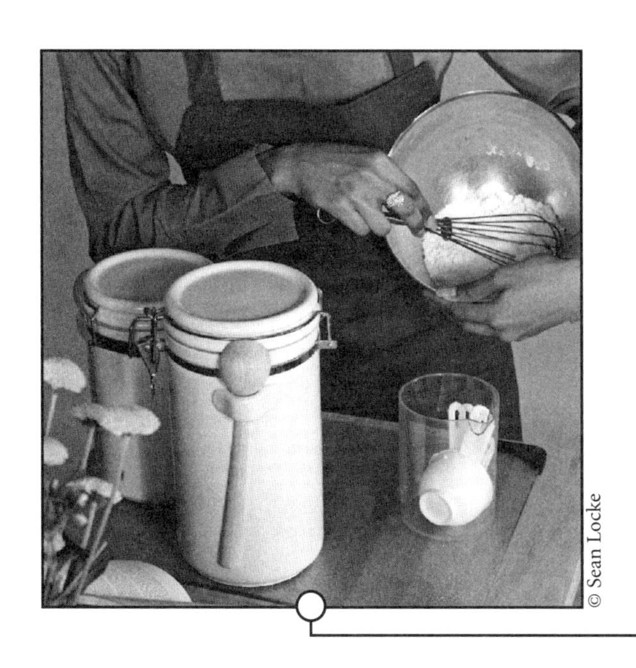

© Sean Locke

What is $2\frac{1}{2}$ times this amount?

When you multiply a whole number by another whole number, the answer, called a **product**, gets bigger (except when you multiply by 1, in which case the product stays the same, or by 0, in which case the answer is 0). Think of how you figure out your pay when you are paid by the hour (say, \$15 an hour). If you work 3 hours, $\$15 \times 3 = \$45$. Work 10 hours, $\$15 \times 10$ is \$150. The money really piles up!

But introduce a fraction of an hour, for example, $\frac{1}{2}$ an hour, and look what happens. The product is less than \$15. ($\$15 \times \frac{1}{2} = \$7\frac{1}{2}$, or $\$15 \times 0.5 = \$7.50$.)

In this lesson, you will explain the ways multiplication of fractions, decimals, and percents is the same as multiplication with whole numbers, and the ways it is different.

Activity 1: Show Me…

1. Show me $12 \times \frac{3}{4}$

   a. Make a representation to communicate what the equation means to you.

   b. Write a problem where you would use $12 \times \frac{3}{4}$ as a way to find the solution.

2. Show me 2.5×12

   a. Make a representation to communicate what the equation means to you.

   b. Write a problem where you would use 2.5×12 as a way to find the solution.

3. Is $\frac{3}{4} \times \frac{6}{5}$ . . .

a. Greater than or less than $\frac{3}{4}$? Explain.

b. Greater than or less than $\frac{6}{5}$? Explain.

4. $\triangle \times \square = 3$

Fill in the boxes. At least one of the numbers must be a fraction.

Write two different equations that would both be true statements.

Activity 2: Headbands

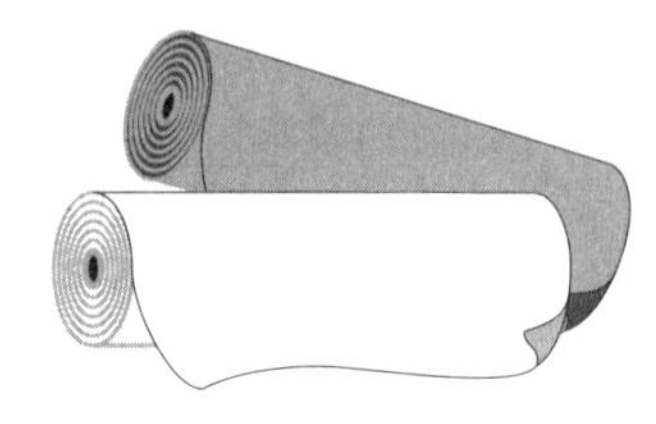

Charlene makes headbands. She is getting ready for the craft fair and wants to make 48 headbands. For each headband she needs $\frac{3}{4}$ yd. of material. She started thinking …

1 headband = $\frac{3}{4}$ yd.

2 headbands = $\frac{3}{4}$ yd. + $\frac{3}{4}$ yd.

1. Continue Charlene's thinking to decide how many yards of material she needs.

2. Peter, Charlene's husband, thought about the problem differently. His thinking went like this:

   "First, I'll give each headband a yard of material—that's a total of 48 yards.

   "That's too much. I'll take forty-eight $\frac{1}{4}$-yards away."

   Continue Peter's reasoning to find out how many yards of material Charlene needs for the headbands.

3. Valerie said, "I don't really like to work with fractions or quarters. I see $\frac{3}{4}$ yd. doubled as 1.5 yd. Then, because I doubled $\frac{3}{4}$, I'll just halve the 48." Show how Valerie's thinking works.

4. Zachary said, "I changed $\frac{3}{4}$ yd. to .75 yd. Since I need 48 of them, I used my calculator and pressed these buttons…"

   What did Zachary enter into his calculator?

5. Show how you would solve this problem.

Activity 3: Paycheck Multiplication

The ABC Company has some good employee benefits. The people in the payroll office use these rules to calculate employees' weekly paychecks.

Payroll Calculation Rules

Multiply the number of hours over 40 by 1.5 to calculate overtime pay.

Multiply the gross pay by 0.035 to calculate Social Security tax (FICA).

Multiply the number of miles traveled by car by 0.425 to calculate mileage reimbursement.

Optional: Multiply the gross pay by 0.1 to calculate the amount going toward retirement.

Optional: Multiply the gross pay by 0.05 to calculate the amount put into a pre-tax medical fund.

John and Mary earn the same hourly rate, $20/hr. John worked a lot this week (57.5 hours) and traveled as well (142 miles). He chooses the retirement option and also contributes to his pre-tax medical fund each week. Mary worked half as many hours as John but traveled four times as many miles. She also chooses the retirement and pre-tax medical options.

1. Fill in a pay stub for Mary and for John.
2. Which one ended up with more take-home pay?

Compare your calculations with a partner.

John H. Smith
3 Main Street, Apt. 2b
Whoville, SC

Earnings Statement

Period ending: 04/02/06
Pay Date: 04/09/06

Earnings	
Gross Pay	
Travel Reimbursement	
Deductions	
Federal Income Tax	280.50
Pre-tax Medical	
Retirement	
FICA	
Net Pay	

Mary J. Brown
5368 68th Street S.W.
Anytown, ME

Earnings Statement

Period ending: 04/02/06
Pay Date: 04/09/06

Earnings	
Gross Pay	
Travel Reimbursement	
Deductions	
Federal Income Tax	150.76
Pre-tax Medical	
Retirement	
FICA	
Net Pay	

3. Choose one of the rules. Show another way to do the calculation.

4. How do the ABC Company's benefits compare with your company's? (If this does not apply to you directly, ask a friend or relative about his or her company's benefits and deductions.)

5. What are some **multipliers** your (or a friend's) company payroll office uses?

Activity 4: Target 1, 10, and 100

Think these problems out with a partner.

Target 1

Regina is practicing some multiplication facts with her 6th-grade nephews, Dan and Martin. She says to Dan, "I'll give you a number. You tell me what we should multiply it by to get 1."

Help Dan out…

1. $\frac{1}{2} \times \square = 1$
2. $4 \times \square = 1$
3. $25 \times \square = 1$
4. $\frac{1}{20} \times \square = 1$
5. $\frac{4}{5} \times \square = 1$

Target 10

Now Dan is trying to get to 10 from these fractions and needs your help.

6. $\frac{1}{2} \times \square = 10$
7. $4 \times \square = 10$
8. $25 \times \square = 10$
9. $\frac{1}{20} \times \square = 10$
10. $\frac{4}{5} \times \square = 10$

Target 100

Martin is working on getting to 100. Help him out.

11. $\frac{1}{2} \times \square = 100$
12. $4 \times \square = 100$
13. $25 \times \square = 100$
14. $\frac{1}{20} \times \square = 100$
15. $\frac{4}{5} \times \square = 100$

16. Look back at the three targets: 1, 10, and 100. What patterns do you notice?

Practice: Seeing Patterns

Do the math in your head or with a calculator. Then look for patterns.

1. a. $56 \times \frac{1}{2} =$ c. $850 \times 0.5 =$ e. $3.5 \times \frac{1}{2} =$

   b. $56 \div 2 =$ d. $850 \div 2 =$ f. $3.5 \div 2 =$

   g. What pattern do you see?

2. a. $56 \times \frac{1}{10} =$ c. $850 \times 0.1 =$ e. $3.5 \times \frac{1}{10} =$

   b. $56 \div 10 =$ d. $850 \div 10 =$ f. $3.5 \div 10 =$

   g. What pattern do you see?

3. Change these multiplication problems into division problems and solve.

   a. $\frac{1}{2} \times \frac{1}{8} =$

   b. $4.5 \times \frac{1}{4} =$

4. Change these division problems into multiplication problems and solve.

   a. $15 \div 10 =$

   b. $7.5 \div 2 =$

5. a. Make up your own multiplication problem.

b. Now change the problem to a division problem.

c. Which did you find easier to solve, the multiplication or the division? Explain.

Practice: Compare...

1. My doctor told me I was eating too many eggs each week. She said, "Eat $\frac{3}{4}$ of the eggs you are now eating." I was eating 12 eggs.

   a. How many eggs am I now allowed to eat?

   b. Compared to what I used to eat, must I now eat fewer or more eggs?

   c. How could you use a calculator to solve this problem?

2. Jennifer's new car uses $\frac{2}{3}$ as much gasoline as her old car did. With her old car, she needed a full tank of gas (18 gallons) to travel 200 miles.

   a. How many gallons would she use with her new car to make the same trip?

   b. Compared to the car she had before, does her new car use less gasoline or more?

   c. How could you use a calculator to solve this problem?

   d. Compare the gas mileage of the cars in miles per gallon.

3. Inez switched jobs and now makes $\frac{5}{6}$ the salary she made in her previous job. Her weekly salary was $240.

   a. How much does Inez make per week now?

   b. Compared to what she earned before, does she now earn less or more?

   c. How could you use a calculator to solve this problem?

4. Elena purchased a sweater on sale. The original price was $48. On sale, she paid 75%.

   a. How much did she pay for the sweater?

   b. Compared to the original price, was the sale price of the sweater less or more?

5. Dennis worked out regularly before he became ill. He biked for 45 minutes each morning. He says he is now biking about $\frac{4}{5}$ of the time he did before becoming ill.

   a. How long does Dennis bike now?

   b. Compared to how much he biked before his illness, is he now biking less or more?

Practice: What Does It Mean?

Show what each of the following problems could mean. Use pictures or words.

1. 6.25×0.5

2. $\frac{3}{4} \times \frac{1}{3}$

3. 1.5×11

Practice: How Much More or Less?

Answer each question. Explain how you arrived at your answers.

1. How much more than 2 is $\frac{4}{5} \times 3$?

2. How much more than 10 is $\frac{3}{10} \times 50$?

3. How much less than $\frac{1}{2}$ is $12 \times \frac{1}{25}$?

4. How close to 1 is $\frac{2}{5} \times 2$?

5. How much more do you need to add to 100 to get $120 \times \frac{2}{3}$?

Calculator Practice: Decimals, Decimals

Two people play this game. Taking turns, look at the Table and the Factor Charts below (one square, one round). Choose any two **factors**, one from each chart. If the product of the two factors is on the Table, you capture the square. The first person to capture four squares on the Table is the winner. Use a calculator to verify your products.

Table

278.4	46.4	60.6	10.56
16	4	76.8	136.4
99.2	63.8	47.04	28.2
14.5	5.5	350.6	2.45

Factor Charts

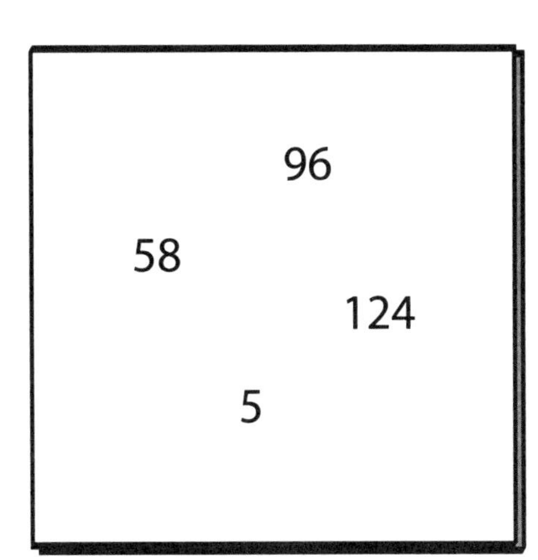

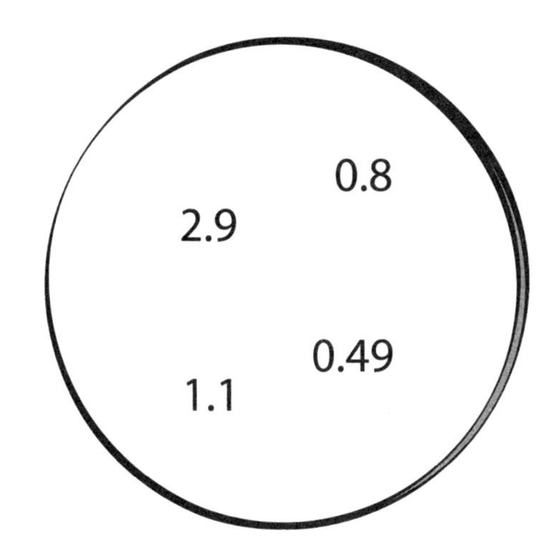

Extension: Guess My Number

1. I'm thinking of a number that when multiplied by $\frac{1}{2}$ is 29. Guess my number.

2. I'm thinking of a fraction that when multiplied by $\frac{19}{20}$ is equal to 1. Guess my number.

3. I'm thinking of a decimal that when multiplied by $\frac{1}{10}$ is 5.2. Guess my number.

4. I'm thinking of a number that when multiplied by 0.1 and then divided by 2 is 100. Guess my number.

Test Practice

Use mental math to choose the best answer for these problems.

1. Lee's old rent was $500 per month. His new rent is $\frac{4}{5}$ of that sum. How much is Lee's new rent?
   (1) $400
   (2) $450
   (3) $540
   (4) $545
   (5) $600

2. Evelyn's phone bill is higher in November and December than during the rest of the year because she calls everyone during the holidays. She says, "I multiply my usual $30 monthly bill by $\frac{1}{6}$, and that's the extra that I pay." How much extra does Evelyn spend in November and December?
   (1) $5
   (2) $6
   (3) $8
   (4) $16
   (5) $18

3. Jonathan changed jobs. It now takes him $\frac{7}{8}$ the time it took him previously to get to work. Instead of 40 minutes, it now takes him how many minutes to get to work?
   (1) $\frac{8}{7} \times 40$ minutes
   (2) 7×40 minutes
   (3) $\frac{7}{8} \times 40$ minutes
   (4) 40 minutes
   (5) 45 minutes

4. Susan is baking brownies. The recipe calls for 4 eggs, but she only has 3. She decides to make $\frac{3}{4}$ the recipe. The original recipe calls for 2 cups of flour. How many cups of flour will she need for the recipe she is making?
   (1) 4 cups
   (2) 3 cups
   (3) 2 cups
   (4) $1\frac{1}{2}$ cups
   (5) $1\frac{1}{4}$ cups

5. What is another way to solve $\frac{1}{4} \times 18$?
   (1) $\frac{1}{18} \times 4$
   (2) $4 \div 18$
   (3) $18 \div 4$
   (4) 4×18
   (5) $\frac{1}{4} \div 18$

6. Rose is saving for a big trip to visit her family in the Philippines. She figures that if she sets aside $\frac{1}{3}$ of her monthly salary for six months, she can make the trip. Her monthly salary is $750. What is the total she is putting aside for her trip?

LESSON 5

Division—Splitting and Sharing

© Oleg Prikhodko

How are we going to split this?

What comes to mind when you think about the division problem \$500 ÷ 4? Do you connect division with the act of sharing an amount or dealing it out so that each of four people gets \$125, an equal amount?

But what does \$500 ÷ 4.3 mean? How would the \$500 be shared then?

In this lesson, you will focus on the effect of division with fractions and decimals. How is division with fractions and decimals the same and how is it different from division with whole numbers?

Activity 1: Weekly Expenses

One bank requires home mortgage applicants to fill out an expense report.

INSTRUCTIONS: All expense figures must be listed by their *weekly* total. *Do not* list expenses by their *monthly* total. To compute the weekly expense, divide the average monthly expense by 4.3. For example, if your average rent is $500.00 per month, divide 500 by 4.3. This will give you a weekly expense of $116.28.

1. Why would the bank tell its mortgage applicants to divide by 4.3?

Christina is very organized. She has kept a record for the past year of all her monthly payments. She will use this information to report her weekly total for each expense.

Category	Jan	Feb	Mar	Apr	May	Jun
Rent	$1,000	$1,000	$1,000	$1,000	$1,000	$1,000
Cell Phone	68.64	156.85	380.53	327.48	108.20	116.78
Heat (oil)	171.99	334.35	0	232.78	0	0
Electricity	27.59	44.26	55.10	57.77	39.79	56.16

Category	Jul	Aug	Sep	Oct	Nov	Dec
Rent	$1,000	$1,000	$1,000	$1,000	$1,000	$1,000
Cell Phone	93.18	201.35	94.09	58.75	115.06	106.54
Heat (oil)	0	0	0	0	0	56.75
Electricity	30.54	33.47	43.12	40.68	26.49	28.82

2. Complete the table.

   a. First, mentally estimate the weekly expense for each category.

   b. Then use the rule in the instructions to get a more exact amount.

Item	Average Monthly Expense	Estimated Weekly Expense	Weekly Expense to Report to the Bank
Rent			
Phone			
Heat			
Electricity			

   c. Describe your estimation (mental math) processes.

   d. Describe how you calculated the exact weekly expense to report.

3. Try this on your calculator: Instead of $500 \div 4.3 =$, enter $4.3 \div 500 =$

a. When you switch the numbers in a division problem, what is the same?

b. What is different?

c. Think of a financial or consumer situation where you might need to calculate $4.3 \div 500$. Write about or diagram that situation.

Activity 2: What Is the Message?

Math symbols—numbers and operation signs—send a message. Sometimes a division symbol sends the message of the action of sharing or splitting an amount. In this activity, you will think about the sharing message in a division problem.

- C. $2.5 \div 2 =$
- B. $5.25 \div 2 =$
- F. $3\frac{1}{2} \div 5 =$
- $1\frac{1}{2} \div 3 =$
- A. $2\frac{1}{2} \div 5 =$
- D. $.25 \div 2.0 =$

With a partner, select one division problem card.

Make the math come alive!

Demonstrate with a diagram or objects what the problem means to you. Then tell a brief story or describe a situation that might correspond to the math symbols. Finally, do the math with a calculator or with paper and pencil. Do your picture, story, and calculation all show the same result?

Prepare a poster of your results to share with the class.

Include the following on the poster:

1. A picture interpretation of the problem
2. A real-life situation (Tell the story.)
3. Your completed division equation

Activity 3: Target Practice 0.1

How could you split a number to get 0.1 in each group?

Can you figure it out in three turns or less on the calculator? Try it. Keep trying!

Round 1: Start with 8. Your goal is to divide 8 by a number to reach 0.1.

First try $8 \div$ _____ = _____

Second try $8 \div$ _____ = _____

Third try $8 \div$ _____ = _____

Round 2: Start with 42.

First try $42 \div$ _____ = _____

Second try $42 \div$ _____ = _____

Third try $42 \div$ _____ = _____

Round 3: Start with 249.6.

First try $249.6 \div$ _____ = _____

Second try $249.6 \div$ _____ = _____

Third try $249.6 \div$ _____ = _____

Round 4: Start with $9\frac{3}{4}$.

First try $9\frac{3}{4} \div$ _____ = _____

Second try $9\frac{3}{4} \div$ _____ = _____

Third try $9\frac{3}{4} \div$ _____ = _____

Round 5: Start with $\frac{7}{8}$.

First try $\frac{7}{8} \div$ _____ = _____

Second try $\frac{7}{8} \div$ _____ = _____

Third try $\frac{7}{8} \div$ _____ = _____

Round 6: Start with 0.08.

First try	0.08 ÷ _____ = _____
Second try	0.08 ÷ _____ = _____
Third try	0.08 ÷ _____ = _____

1. What patterns do you see?

2. Start with a really crazy number. What must you divide by to get an answer of 0.1? Write the problem and the solution.

3. Now return to the division problems from Rounds 1–6, this time with a target of 0.01. What number would you divide by to get 0.01 in each group? Look for a pattern.

a. 8 ÷ _____ = 0.01

b. 42 ÷ _____ = 0.01

c. 249.6 ÷ _____ = 0.01

d. $9\frac{3}{4}$ ÷ _____ = 0.01

e. $\frac{7}{8}$ ÷ _____ = 0.01

f. 0.08 ÷ _____ = 0.01

Practice: Four Ways to Write Division

Division can be recorded in several ways.

Be careful. The order of the numbers makes a difference in division.

Complete the table. The first problem has been done as an example.

1. $8\overline{)56}$	$56 \div 8$	$\frac{56}{8}$	56 divided by 8
2.	$0.6 \div 8$		
3. $7.2\overline{)1.24}$			
4.		$\frac{0.8}{4}$	
5.	$1\frac{1}{2} \div 3$		
6.			$\frac{1}{4}$ divided by 10
7.			20.3 divided by 10
8. $10.3\overline{)1}$			

Practice: Which Is Not the Same?

Order makes a difference in division. $\frac{1}{2} \div 5$ is not the same as $5 \div \frac{1}{2}$.

Use what you know about division notation and rules of order to choose the one expression that is not the same as the others.

(Reminder: Always do the math in parentheses first.)

1. **a.** $2.8 \div 7$

b. 2.8 divided by 7

c. $2.8\overline{)7}$

d. $\frac{2.8}{7}$

2. **a.** $\frac{1.5}{1,000}$

b. $1,000 \div 1\frac{1}{2}$

c. $1.5\overline{)1,000}$

d. $\frac{1,000}{1\frac{1}{2}}$

3. **a.** 10 divided by $\frac{1}{6}$

b. $(14 - 4) \div \frac{1}{6}$

c. $10\overline{)0.6}$

d. $\frac{10}{0.167}$

4. **a.** $(72.8 + 7.2) \div 4$

b. $72.8 + (7.2 \div 4)$

c. $(72.8 \div 4) + (7.2 \div 4)$

d. $18.2 + 1.8$

5. a. $\frac{1}{2} \div \frac{1}{4}$

b. $0.25\overline{)0.5}$

c. $0.5\overline{)0.25}$

d. $\frac{0.5}{0.25}$

6. a. $\frac{2\frac{7}{8}}{3}$

b. $2.875 \div 3$

c. $\frac{(2 + \frac{7}{8})}{3}$

d. $2.875\overline{)\ 3\ }$

Practice: Show Me...

First, do the math.

Then make up a situation or draw a picture that matches the math.

1. a. $4.5 \div 3 =$

   b. A picture to match the math:

2. a. $1.2 \div 5 =$

   b. A picture to match the math:

3. a. $\$12.5 \div 100 =$

   b. A situation that matches the division problem:

4. a. $100 \div 12.5 =$

   b. A situation that matches the division problem:

Practice: Where's the Point?

Use your understanding of division as sharing to place the decimal point in the correct spot in the underlined number.

1. $234.6 \div 22.58 = \underline{1\ 0\ 3\ 8\ 9\ 7\ 2\ 4}$

2. $2346.0 \div \underline{2\ 2\ 5\ 8} = 10.389724$

3. $234.6 \div 2.258 = \underline{1\ 0\ 3\ 8\ 9\ 7\ 2\ 4}$

4. $23.46 \div 22.58 = \underline{1\ 0\ 3\ 8\ 9\ 7\ 2\ 4}$

5. $6.5 \div 10 = \underline{00650}$

6. $6.5 \div 100 = \underline{00650}$

7. $6.5 \div \underline{100} = 6.5$

8. $15\frac{3}{4} \div 2.8 = \underline{5625}$

9. $1.575 \div 2.8 = \underline{5625}$

10. $\underline{1575} \div 28 = 0.5625$

Practice: Multiplication and Division Patterns

1. Fill in the chart. Check your work with a calculator. The first problem has been done for you.

Original Amount	Multiplied by 2	Divided by 2	Multiplied by $\frac{1}{2}$	Divided by $\frac{1}{2}$
a. 7	$7 \times 2 = 14$	$7 \div 2 = 3\frac{1}{2}$	$7 \times \frac{1}{2} = 3\frac{1}{2}$	$7 \div \frac{1}{2} = 14$
b. 26				
c. 0.5				
d. $12\frac{1}{2}$				
e. 8.9				

2. Fill in the chart. Check your work with a calculator. The first problem has been done for you.

Original Amount	Multiplied by 4	Divided by 4	Multiplied by $\frac{1}{4}$	Divided by $\frac{1}{4}$
a. 7	$7 \times 4 = 28$	$7 \div 4 = 1\frac{3}{4}$	$7 \times \frac{1}{4} = 1\frac{3}{4}$	$7 \div \frac{1}{4} = 28$
b. 26				
c. 0.5				
d. $12\frac{1}{2}$				
e. 8.9				

3. What patterns do you notice?

Practice: Division Patterns

1. Continue the pattern.

$10 \div 5 = 2$

$9 \div 5 = 1.8$

$8 \div 5 =$

$7 \div 5 =$

$6 \div 5 =$

$5 \div 5 =$

$4 \div 5 =$

$3 \div 5 =$

$2 \div 5 =$

$1 \div 5 =$

2. What pattern do you notice? Why does it work?

3. Continue the pattern.

$10 \div 2.5 = 4$

$9 \div 2.5 = 3.6$

$8 \div 2.5 =$

$7 \div 2.5 =$

$6 \div 2.5 =$

$5 \div 2.5 =$

$4 \div 2.5 =$

$3 \div 2.5 =$

$2 \div =2.5 =$

$1 \div 2.5 =$

4. What pattern do you notice? Why does it work?

Calculator Practice: Fraction and Decimal Division

If you can do the calculation in your head, do so. Otherwise, use a calculator to get an exact answer. In all cases, estimate first!

The Problem	A Thoughtful Estimate	The Exact Answer
1. $1.11\overline{)37.74}$		
2. $0.9516 \div 4$		
3. $\frac{38.6}{2}$		
4. $199 \div 2.3$		
5. $16.8 + 2$		
6. $16.8 - 2$		
7. 16.8×2		
8. $16.8 \div 2$		
9. $6\frac{3}{5} \times \frac{1}{10}$		
10. $6\frac{3}{5} \div 10$		

Extension: Mirror Frames

Three mirrors are pictured below. Determine how much framing is needed to go around each mirror. Each mirror has all sides equal.

1. 

Side = 11.25″

____

2. 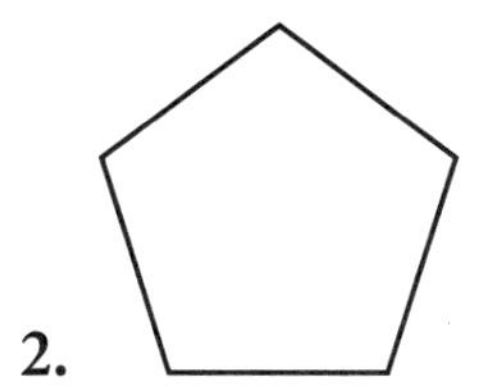

Side = 6.9″

____

3. 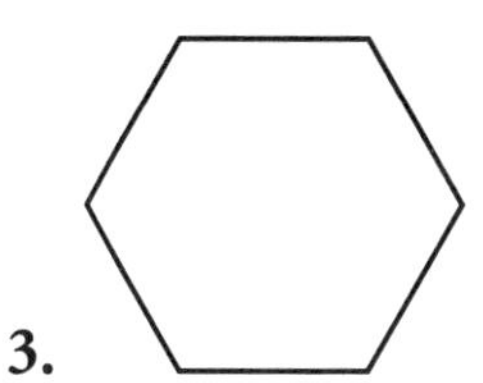

Side = $20\frac{3}{8}$″

____

Here are three more mirrors with equal sides. The amount of total framing is given for each. What is the length of one side for each mirror?

4. 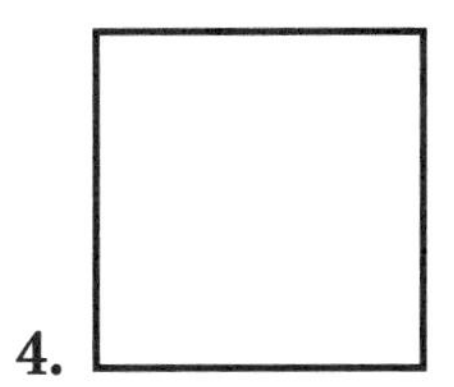

Framing = 11.25″

____

5. 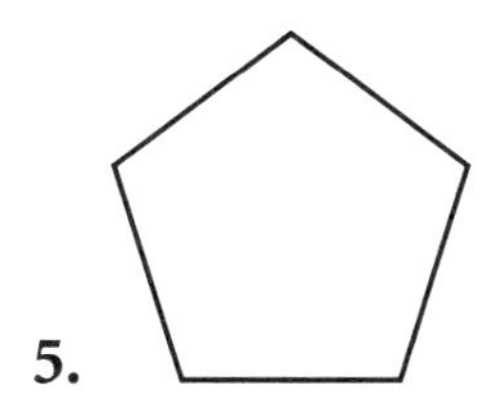

Framing = 6.9″

____

6. 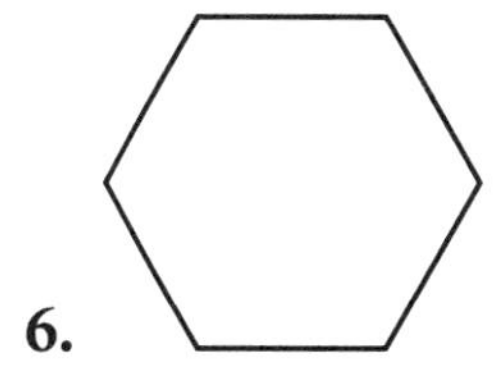

Framing = $20\frac{3}{8}$″

____

7. If it takes 30″ of framing to go around an octagonal mirror, what is the length of each side? All sides are equal.

Test Practice

1. I saved $9,000 over a period of five-and-a-half years. If I saved the same amount each year, how much did I save annually? Which of the following expressions could match the story?

(1) $9{,}000 \times 5.5$

(2) $5.5\overline{)9{,}000}$

(3) $5\frac{1}{2} \div 9{,}000$

(4) $\frac{9{,}000}{12}$

(5) $5.2 \times 9{,}000$

2. The rainfall for January, February, and March was 2.8, 3.55, and 0.4 inches, respectively. What was the average rainfall per month?

(1) 0.49 inches

(2) 2.25 inches

(3) 2.8 inches

(4) 3.55 inches

(5) 6.75 inches

3. $\frac{1}{2} \div 2 =$

(1) 0.1

(2) 0.25

(3) 0.4

(4) 1.0

(5) 2.5

4. What is 2 divided by 5?

(1) 0.1

(2) 0.25

(3) 0.4

(4) 1.0

(5) 2.5

5. Which of the following numbers is closest in value to $\frac{15}{16}$?

(1) 0.2

(2) 0.4

(3) 0.6

(4) 0.8

(5) 1.0

6. What number belongs in the box?

$4.56 \div \square = 0.1$

LESSON 6

Division—How Many ____ in ____?

© Steven Lunetta Photography

How many $\frac{3}{4}''$ slices from this cake?

When faced with a division problem such as 12 ÷ 5, you may think of it in one of two ways. If you think of division as a **splitting** action, you might picture $12 shared among 5 people. The result would be $2.40 per person.

But you could also look at it another way. You might think, "**How many** $5 are **in** $12?" The answer is more than 2, but not quite 3. In fact, the answer is 2.4.

In the last lesson, you looked at the operation of division as a way to split up an amount into equal parts. You interpreted $500 ÷ 4.3 as 500 dollars split among 4.3 weeks in order to find out how much money was accounted for in each week. In this lesson, you will be asked to take a different look at division. When you see 500 ÷ 4.3, you could also ask yourself, "How many $4.30 amounts are in $500?"

As you work with objects, diagrams, and measurements, keep in mind also the ways the pictures of division and multiplication relate.

Activity 1: Show Me…

1. Show me how many groups of three cans can be made from a case of 24 soda cans.

   a. Make a representation to communicate what that means to you.

   b. Write the problem using math symbols.

2. Andy won the lottery for 1.6 million dollars. He needs to share his winnings with five friends who chip in every week for the ticket, so he decides to give each one 0.25 million dollars and keep the rest for himself. Is this fair? Show me.

   a. Make a representation to communicate what that means to you.

   b. Write the problem using math symbols.

3. Show me how many $\frac{1}{10}$'s are in $\frac{3}{5}$.

   a. Make a representation to communicate what that means to you.

   b. Write the problem using math symbols.

4. Show me how many $\frac{1}{2}$'s are in $\frac{1}{8}$.

   a. Make a representation to communicate what that means to you.

   b. Write the problem using math symbols.

5. In Problems 1–4, what connections to multiplication do you see?

Activity 2: Pattern Block Division

Use pattern blocks to explore these questions.

1. For this activity the yellow hexagon will have a value of 1. What is the value of

   a. The green triangle?

   b. The red trapezoid?

   c. The blue parallelogram?

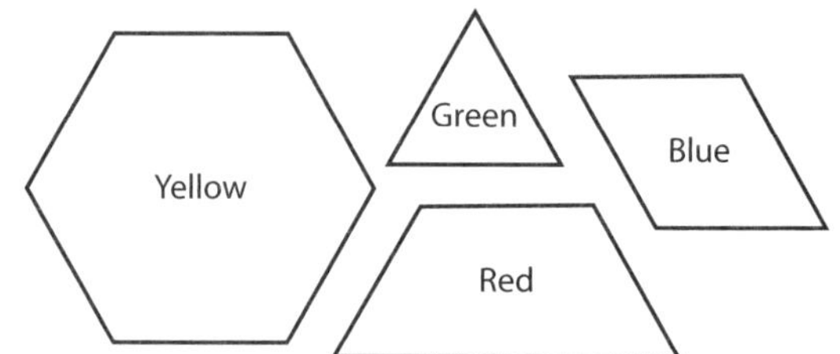

Answer the following division questions by drawing a picture, writing the math symbols, and performing the division on a calculator. Then show another way to explain the answer.

An example has been done for you:

How many triangles are in 2 hexagons?

Answer: 12

a. A picture:

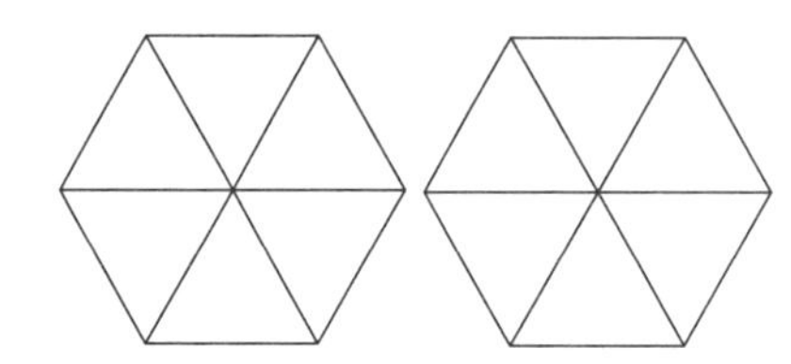

b. Math symbols: $2 \div \frac{1}{6} = 12$

c. Calculator: Convert $\frac{1}{6}$ to a decimal 0.167 (rounded)

$2 \div 0.167 = 11.97$, close to 12

d. Show another way to explain the answer.

$1 \div \frac{1}{6} = 6$, so $2 \times 6 = 12$

2. How many parallelograms are in a trapezoid?

   Answer: ____

   a. A picture:

b. Math symbols:

c. Calculator:

d. Show another way to explain the answer.

3. How many hexagons are in two triangles?

Answer: ____

a. A picture:

b. Math symbols:

c. Calculator:

d. Show another way to explain the answer.

4. How many trapezoids are in a parallelogram?

Answer: ____

a. A picture:

b. Math symbols:

c. Calculator:

d. Show another way to explain the answer.

5. Now use the shapes to *ask and answer* your own "How many ___ in ___?" question. Show the answer with a picture, in math symbols, and with a calculator.

Answer: ____

a. A picture:

b. Math symbols:

c. Calculator:

d. Show another way to explain the answer.

Activity 3: Making Do

Carla loves to cook and has a new recipe for turkey meatloaf. Unfortunately, when she visits her boyfriend, his kitchen is not as well equipped as hers. As she searches through the drawers, looking for measuring spoons and cups, all she can find are the $\frac{1}{4}$-teaspoon (tsp.) and the $\frac{1}{3}$-cup (c.) measures.

Turkey Meatloaf
1 tsp. olive oil
1 1/2 tsp. Worcestershire sauce
1 c. fine fresh bread crumbs
1 tablespoon (T.) minced garlic
1/4 c. plus 1 T. ketchup
1 1/2 c. finely chopped onion
✓ 1 medium carrot, cut into 1/8-inch dice
✓ 3/4 lb. cremini mushrooms, trimmed and very finely chopped in a food processor
✓ Salt and pepper to taste
✓ 1/3 c. finely chopped fresh parsley
✓ 1/3 c. 1% milk
✓ 1 whole large egg, lightly beaten
✓ 1 large egg white, lightly beaten
✓ 1 1/4 lb. ground turkey (mix of dark and light meat)

Carla checked off the ingredients she knows how to measure, but she's not sure what to do about the rest. To keep track of measuring, she made a chart.

1. Help Carla fill in the chart. Decide whether she should use the $\frac{1}{3}$-cup measure or the $\frac{1}{4}$-tsp. measure, and how many of each measure she needs for the different ingredients.

Ingredient	How many $\frac{1}{3}$-cup Measures?	How many $\frac{1}{4}$-tsp. Measures?
a. 1 tsp. olive oil		
b. $1\frac{1}{2}$ tsp. Worcestershire sauce		
c. 1 c. bread crumbs		
d. 1 T. minced garlic		
e. $\frac{1}{4}$ c. plus 1 T. ketchup		
f. $1\frac{1}{2}$ c. finely chopped onion		

2. Explain how you figured out how to measure the **Worcestershire sauce.**

a. Draw a picture.

b. Write it using math symbols to show the division.

c. Do the division on a calculator.

3. Explain how you figured out how to measure the **onions.**

a. Draw a picture.

b. Write it using math symbols to show the division.

c. Do the division on a calculator.

Activity 4: Target 100

This is game of finding patterns as you ask the division question "How many ___ in ___?"

With a partner, take turns until you get to 100. Use a calculator to verify your answers.

1. 500 ÷ ☐ = 100
2. 250 ÷ ☐ = 100
3. 50 ÷ ☐ =100
4. 10 ÷ ☐ = 100
5. 1.5 ÷ ☐ = 100
6. 0.5 ÷ ☐ = 100
7. 0.1 ÷ ☐ = 100
8. Do you see a pattern? Explain.

Practice: On a Diet

You are on a diet and may eat $\frac{1}{3}$ pound of turkey each day.

Your butcher gives you three equal slices weighing a total of $\frac{3}{4}$ pound. So, how many slices can you eat that day?

1. Show a mathematical solution.

2. Show the solution with a drawing.

Practice: Can You See It?

Solve each problem with a drawing:

1. $1\frac{1}{2} \div 2 =$

2. $1\frac{1}{2} \times \frac{1}{2} =$

3. $1\frac{1}{2}$ divided by $\frac{3}{5}$

4. $1\frac{1}{2}$ multiplied by $\frac{5}{3}$

5. What do you notice?

Practice: Think Metric

1. How many 2.5 cm segments are in a 10 cm segment?
   a. Division equation:

   b. Picture or number line:

2. How many 0.1 liter portions are in 3 liters?
   a. Division equation:

   b. Picture or number line:

3. How many 3.5 kg portions are in 16 kg?
   a. Division equation:

   b. Picture or number line:

4. How many 0.75 km are in 5 km?
   a. Division equation:

   b. Picture or number line:

Calculator Practice: A Mixed Bag

Solve these problems with a calculator. Remember the rules for order of operations, always doing the math that is inside the parentheses first.

1. (2.6×4)

2. $\dfrac{(3\frac{1}{2} + 2\frac{1}{2})}{6.5}$

3. $6.8(2.65 + 3.5)$

4. $\dfrac{(5 - 2\frac{1}{5})}{1.25}$

Extension: Geometric Formulas

All diagrams are drawn to scale.

First, estimate the answer by "eyeballing" the picture. Then use the formula (and a calculator if you wish) to get an exact answer.

1. The distance around the outside of this **circle**, the **circumference**, is 61.6 inches.

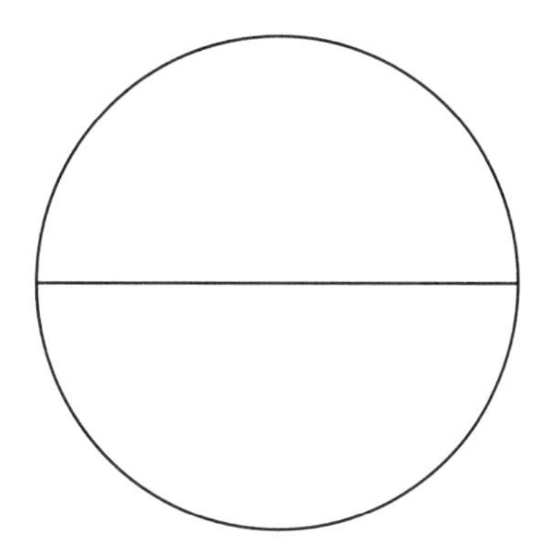

   a. Estimate the distance across the circle, the **diameter**, in inches.

   b. Calculate the diameter using the formula: Diameter = Circumference ÷ 3.14

2. The total distance around the outside of this **rectangle** is 20.5 cm. The width is 4.2 cm.

   a. Estimate the length of the rectangle in cm.

   b. Calculate the length using the formula: $L = (P - 2w) \div 2$.

Test Practice

1. $2.5 \div 0.25 =$

   (1) 0.1

   (2) 0.625

   (3) 1.0

   (4) 10

   (5) 100

2. $\frac{4}{5} \div \frac{4}{5} =$

   (1) 0

   (2) $\frac{16}{25}$

   (3) $\frac{8}{10}$

   (4) 1

   (5) $1\frac{1}{4}$

3. $20 \div \square = 100$

   (1) 0.02

   (2) 0. 2

   (3) 0.5

   (4) 2

   (5) 5

4. Which of the following will result in an answer that is not the same as the others?

   (1) $100 \times \frac{1}{5}$

   (2) $100 \div 5$

   (3) $(100 \times \frac{1}{10}) \times 2$

   (4) $\frac{100}{5} \times \frac{1}{5}$

   (5) 100×0.2

5. Will had $\frac{2}{3}$ of a can of frosting to frost $\frac{1}{2}$ a cake. How much of the cake could he frost with the whole can of frosting?

   (1) $\frac{2}{3}$ of the cake

   (2) $\frac{3}{4}$ of the cake

   (3) Two halves of the cake

   (4) The whole cake

   (5) $1\frac{1}{3}$ cakes

6. Rachel has 10 pounds of flour. She is making bread, and each loaf calls for $3\frac{1}{4}$ cups of flour. How many loaves of bread will Rachel be able to make? (1 pound of flour is about 4 cups.)

LESSON 7

Mixing It Up

© Malcolm Romain

What good is a broken calculator?

You have connected many different meanings and interpretations to the operations of addition, subtraction, multiplication, and division. How are the four operations the same? How are they different? How are they related to one another?

In this lesson, you will review what you have learned about using operations with fractions, decimals, and percents. Keep in mind the meaning of the operations as well as the procedures you use.

Activity 1: Going Places with the Four Operations

The chart below shows the distances from Jill's home to four other places where she often travels.

Places Jill Goes	Distance from Jill's Home
Supermarket	0.5 mile
Gym	$\frac{3}{8}$ mile
The Hospital (where she works)	9.1 miles
College	$6\frac{1}{4}$ miles

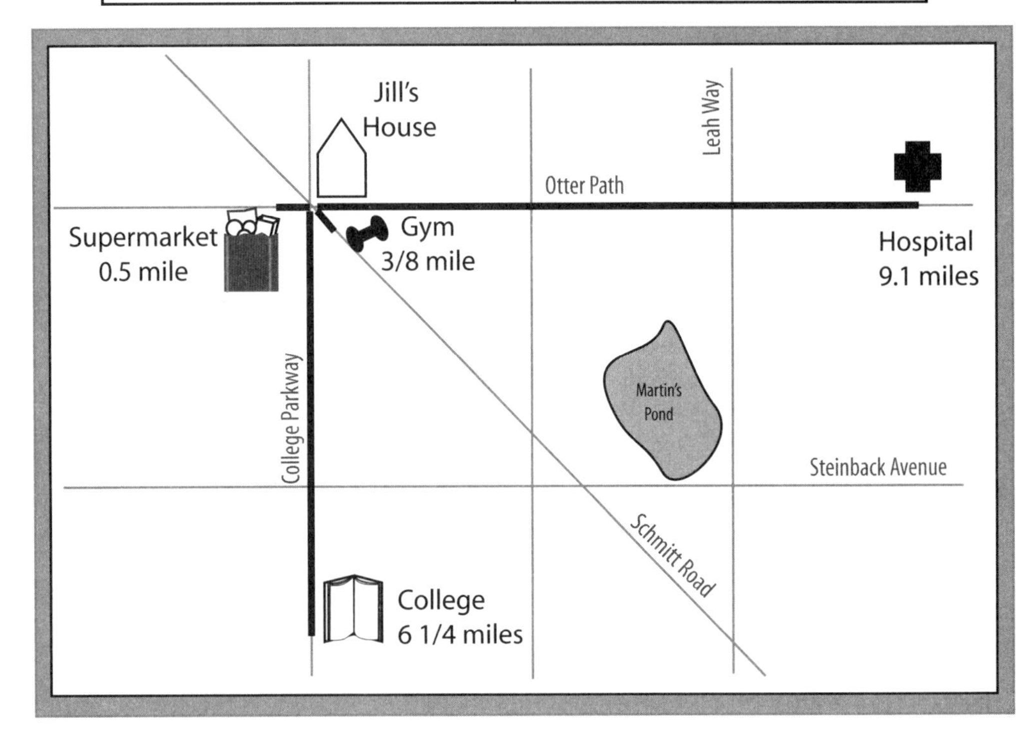

Use some of these distances and the map to create story problems.

1. Think about what you know about **addition** with fractions or decimals.

   a. Write a story problem about Jill's travels where it would make sense to use addition to solve the problem.

b. What is the answer?

c. What do you need to pay attention to when adding fractions or decimals?

2. Think about what you know about **subtraction** with fractions or decimals.

   a. Write a story problem about Jill's travels where it would make sense to use subtraction to solve the problem.

   b. What is the answer?

   c. What do you need to pay attention to when subtracting fractions or decimals?

3. Think about what you know about **multiplication** with fractions or decimals.

   a. Write a story problem about Jill's travels where it would make sense to use multiplication to solve the problem.

   b. What is the answer?

   c. What do you need to pay attention to when multiplying fractions or decimals?

4. Think about what you know about division with fractions or decimals.

   a. Write a story problem about Jill's travels where it would make sense to use division to solve the problem.

b. What is the answer?

c. What do you need to pay attention to when multiplying fractions or decimals?

5. Pick one of the story problems you wrote that could be solved by using a different operation.

a. Show how you could solve it with that alternative operation.

b. Explain the relationship between the two operations.

Activity 2: Broken Calculators

Work with a partner on this activity. Each of you should have a calculator that works. However, you will pretend one of the keys on your calculator is broken.

What would you do if the multiplication key on your calculator were broken?

1. Find 15 × 4 without using the  key.

   Write down the key sequence you pressed.

2. Calculate 1.9 × 6

   Write down the key sequence you pressed.

3. Find $2\frac{3}{8} \times \frac{1}{8}$

   Write down the key sequence you pressed

Trade your paper with a partner. Compare your methods. Do you each arrive at the correct answer using the other's key sequences? Discuss how the sequences are the same, how they are different, and why your methods do or do not work.

What would you do if the division key (÷) on your calculator were broken?

4. Find 15 ÷ 4

Write down the key sequence you pressed.

5. Calculate 7.2 ÷ 10

Write down the key sequence you pressed.

6. Find 956.45 ÷ 100

Write down the key sequence you pressed.

Trade your paper with a partner. Compare your methods. Do you each arrive at the correct answer using the other's key sequences? Discuss how the sequences are the same, how they are different, and why your methods do or do not work.

Activity 3: Number of the Day, 0.1

The number for today is one-tenth.

On the lines below, write as many expressions as you can whose answers equal the number for today. Use as many math symbols and expressions as you can.

You must include at least one fraction, decimal, or percent in each expression. Also, try writing it with some combinations of the symbols $+$, $-$, $\times$, and $\div$.

Practice: Room Measurements

Solve the problems using the geometric formulas and the values provided. Use a calculator, but always estimate first.

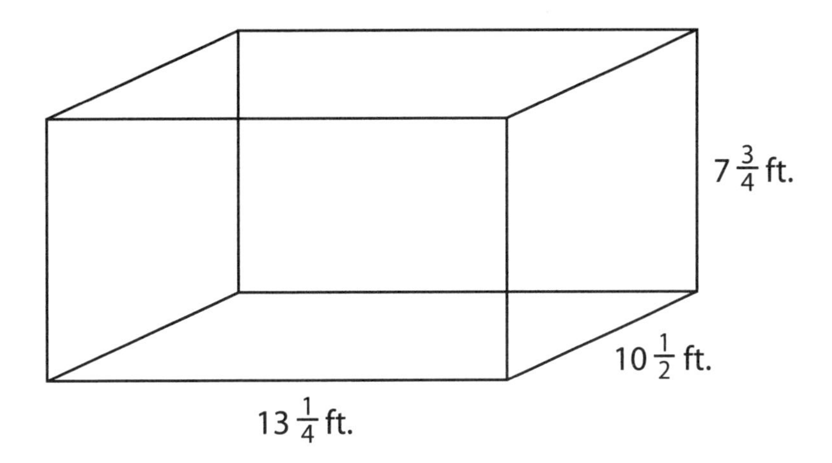

1. You can use this formula to find the **area** of a rectangle:

   Area = length × width. Find the area of the floor of a room that is $10\frac{1}{2}$ feet by $13\frac{1}{4}$ feet.

   a. How I estimated:

   b. How I calculated the answer:

2. The formula to find the **perimeter** of a rectangle is 2 × length + 2 × width. What is the perimeter of the room?

   a. How I estimated:

   b. How I calculated the answer:

3. The formula to find the **volume** of a rectangular solid (a box) is length $\times$ width $\times$ height. Find the volume of air in the room when the height of the ceiling is $7\frac{3}{4}$ feet.

a. How I estimated:

b. How I calculated the answer:

Practice: Running Track Measures

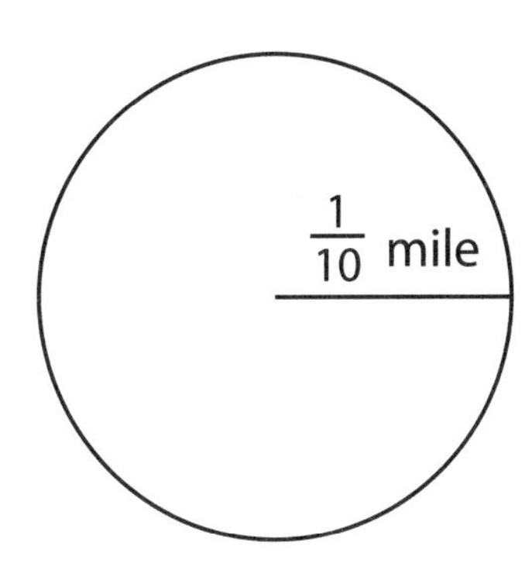

1. The formula to find the circumference (C), the distance around a circle, is $2\pi r$. Find the *distance once around the running track* when $r = \frac{1}{10}$ of a mile. Use $\pi = 3.14$.

a. How I estimated:

b. How I calculated the answer:

2. If a runner did 10 laps around the track, how far did she run?

a. How I estimated:

b. How I calculated the answer:

3. How many times would she run around the track to do a total of 10 miles?

   a. How I estimated:

   b. How I calculated the answer:

Practice: Squaring Fractions and Decimals

Multiplying a number by itself is called "squaring." For example, "5 squared" means 5 × 5. You can use an **exponent**—a raised number—to write the same thing.

$$5^2 = 5 \times 5 = 25$$

Square these fractions or decimals:

1. $(\frac{1}{2})^2 =$

2. $.05^2 =$

3. $(9\frac{1}{3})^2 =$

4. $(2.5)^2 =$

5. $(.2)^2 =$

6. $(.02)^2 =$

7. $(5.8)^2 =$

8. $(\frac{5}{6})^2 =$

Complete the following sentences:

9. When you square a number less than 1, the result is always

____

____

10. When you square a number greater than 1, the result is always

____

____

11. What other patterns do you notice?

Test Practice

First, try these problems in your head. Then check your answers with a calculator.

1. Which amount does not equal 0.5?

(1) $20 \div 40$

(2) 20×0.25

(3) $17.8 - 17.3$

(4) $0.06 + 0.44$

(5) $40 \div 20$

2. $1.5^2 =$

(1) 1.52

(2) 2.25

(3) 3.00

(4) 22.5

(5) 30.0

3. When you multiply 0.01 by 0.25, the answer has a value

(1) Less than 0.01.

(2) Greater than 0.25.

(3) The same as 0.1×2.5.

(4) Between 0.01 and 0.25.

(5) Equal to $\frac{1}{2}$ of 0.1.

4. When you add $0.01 + 0.25$, the answer has a value

(1) Less than 0.01.

(2) Greater than 0.25.

(3) The same as $0.1 + 2.5$.

(4) Between 0.01 and 0.25.

(5) Equal to 0.35.

5. When you divide 0.25 by 0.01, the answer has a value

(1) Equal to $\frac{1}{4} \times \frac{1}{100}$.

(2) Equal to $0.01 \div 0.25$.

(3) Greater than 20.

(4) Less than 20.

(5) The same as 25×100.

6. What is the area in square inches of an $8\frac{1}{2}$″-by-11″ piece of paper?

Closing the Unit: Putting It Together

How would you redesign these packages?

The* Operation Sense *unit is almost finished, and now you have a chance to pull together some of the main ideas that were covered.

For the first activity, you will organize a review of the unit with a *Mind Map*. What did you learn that you think is important? What lessons do you remember most vividly?

Next, you will work on a mini-project, *Cereal Box Math*, to apply what you know to a practical and creative problem. Finally, there's a written assessment with two tasks:

1. Four Problems
2. Working Women

Use any and all of the tools at your disposal to help you think. Draw pictures, use a number line, a tried-and true procedure, or a calculator, supported by estimation. Keep your mind on the prize—to have a personal command of the four operations with fractions, decimals, and percents.

Activity 1: Mind Map

Make a *Mind Map* to organize your understanding of each of the four operations with fractions, decimals, and percents. Use words, pictures, and numbers to show ideas, examples, and things you want to remember. Link the ideas and words that are related. To jog your memory, go back over your notes and work you have done in this unit.

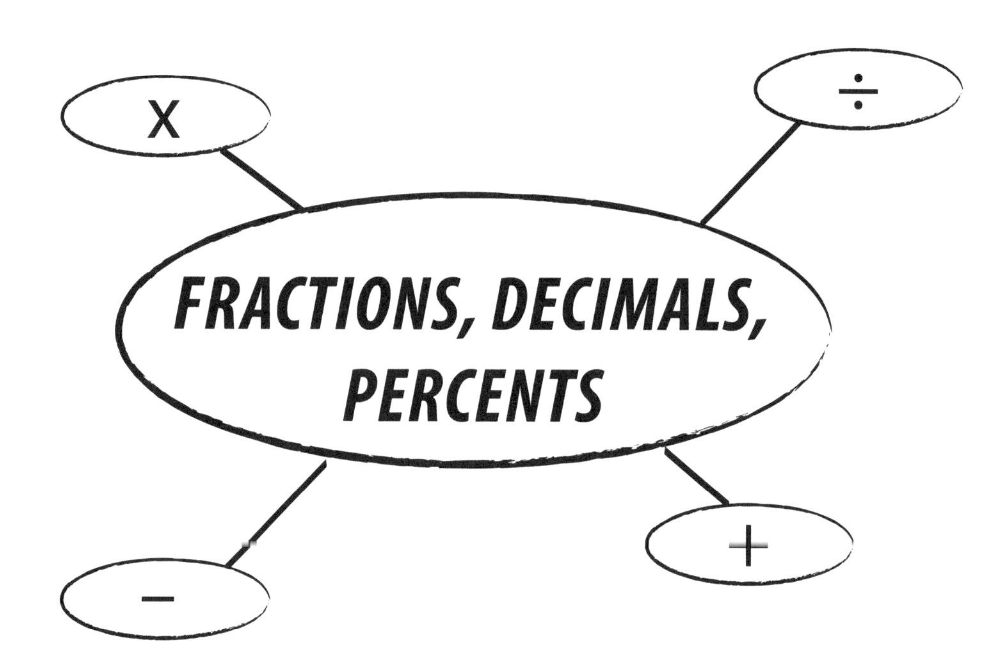

Activity 2 : The Cereal Box Math Project

Materials you will need:

- An empty cereal box
- A ruler marked in cm and mm
- A ruler marked in inches and fractions ($\frac{1}{16}$'s) of an inch
- A calculator

In this project, you and your partner will use your knowledge of fractions and decimals to examine and measure a manufactured cereal box. Then you will be asked to redesign it to meet certain specifications. Prepare a presentation of your redesign.

Using a ruler marked in metric units (cm and mm) and one marked in U.S. customary units, take some measurements.

Heads Up!

Report your metric measurements to the nearest 0.1 cm.

Example:

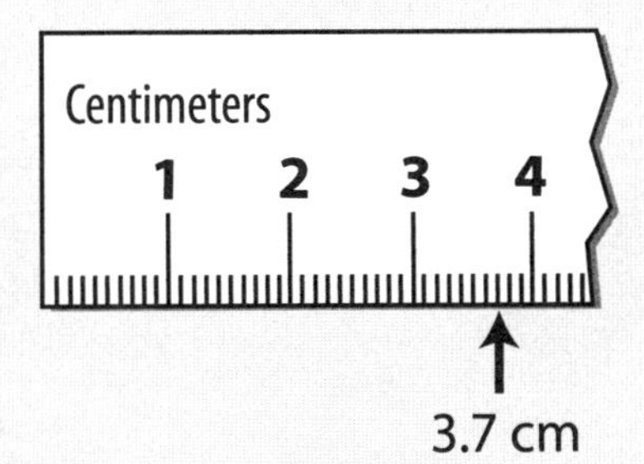

Report your U.S. customary measurements to the nearest 1/8 in.

Example:

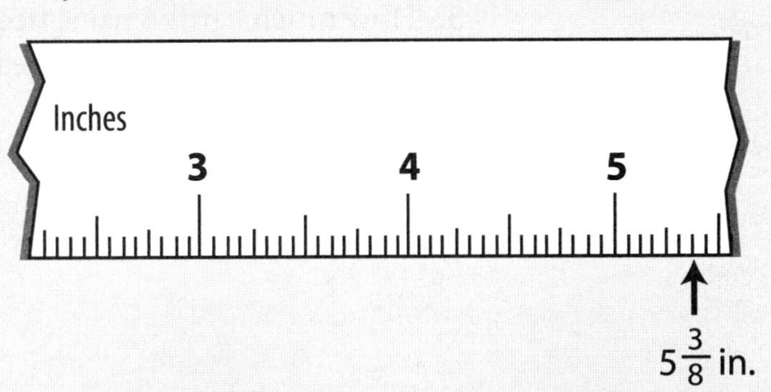

Record your measurements here.

Dimension	Metric Measure (cm)	U.S. Customary Measure (in.)
1. Height	a.	a.
2. Width	b.	b.
3. Depth	c.	c.

Show all work.

4. How much cereal could fit in this box if it were completely filled? In other words, what is the volume of the box? Decide whether you would prefer to work in metric or U.S. customary units.

Use the formula: Volume = length × width × height

5. How much cardboard is needed to make the box? In other words, what is the total surface area?

6. Design another cereal box that will hold the same amount of cereal, but is only half as tall as the original. Draw a picture of the box and label all the dimensions. Explain how you know that the two volumes are equal.

Prepare your presentation. Make sure to include the following information:

- Units you are working with (metric or U.S. customary)
- Dimensions of the original box (height, width, depth)
- Volume of the original box
- Surface area of the original box
- Dimensions of your redesigned box (height, width, depth)
- Volume of your redesigned box
- What was the math? Explain how the volume is still the same, even though the dimensions are not.

VOCABULARY

Lesson	Terms, Symbols, Concepts	Definitions and Examples
Opening the Unit	benchmark	
	decimal	
	fraction	
	percent	
1	conversion	
	denominator	
	digit	
	equivalent	
	numerator	
2	equation	
	hexagon	
	parallelogram	
	trapezoid	
	triangle	
3	common denominator	
	number line	

VOCABULARY *(continued)*

LESSON	TERMS, SYMBOLS, CONCEPTS	DEFINITIONS AND EXAMPLES
4	factor	
	multiplier	
	product	
6	circle	
	circumference	
	diameter	
	rectangle	
7	area	
	exponent	
	perimeter	
	volume	

REFLECTIONS

OPENING THE UNIT: Operation Sense

What problems did you know how to solve? What challenged you?

LESSON 1: Equivalents

Explain the strategies you use to compare the values of any two fractions or decimals. Give some examples.

LESSON 2: Addition—Combining

How did pattern blocks or drawing help you to see fraction addition?

Describe your strategy for adding fractions.

Describe your strategy for adding decimals.

LESSON 3: Subtraction—Take Away, Comparison, and Difference

Write two things you learned about subtraction of fractions and two things you learned about subtraction of decimals.

How are subtraction and addition related?

LESSON 4: Multiplication—Repeated Addition and Portions of Amounts

Multiplication can sometimes be seen as repeated addition. Give an example, using fractions or decimals.

Multiplication can sometimes be seen as finding a portion of an amount. Give an example, using fractions or decimals.

LESSON 5: Division—Splitting and Sharing

Describe an example of what you understand division by sharing or splitting to mean.

Explain why $5 \div 2$ is the same as $5 \times \frac{1}{2}$.

LESSON 6: Division—How Many __ in __?

Write about the difference between division as splitting or dealing out and division as how many __ in __?

LESSON 7: Mixing It Up

What did you learn when you used a broken calculator?

How are multiplication and division related?

CLOSING THE UNIT: Putting It Together

What are the most important ideas and skills you have learned in this unit?

What are you best at?

Where would you like to improve?